PHYSICS TEACHER'S GUIDE

Effective Classroom Demonstrations and Activities

David Kutliroff

PARKER PUBLISHING COMPANY, INC.
WEST NYACK, N.Y.

LIBRARY OF CONGRESS
CATALOG CARD NUMBER: 73-90265

PRINTED IN THE UNITED STATES OF AMERICA
13-674309-9 B&P

To Millicent, my wife, whose patience and assistance made the writing of this book much more pleasant than it would otherwise have been

Toward More Effective Teaching of High School Physics

THIS COLLECTION OF PRESENTATIONS for physics teachers was motivated by eight years of contact with physics teachers at summer institutes.

As a member of the staff of these summer institutes for physics teachers, and sometimes the only high school teacher on the staff, I was often asked, "How do you get this across to your students?"

Often some of the participants contributed suggested presentations, and I have learned much from them.

This book, therefore, was written to present ideas for lessons which will help new teachers get started and to promote the sharing of ideas with my experienced colleagues. Perhaps it will offer alternate methods for presenting familiar lessons.

The choice of topics and the order of presentation is, with some modifications, influenced by the PSSC approach to the teaching of high school physics. It is hoped, however, that much of what is recorded here will be useful to all teachers of physics since what is presented is, as much as we could make it, sound physics. This is applicable in any approach to teaching physics.

The presentations illustrated in this text have evolved through many years of teaching. Much of what is presented in most textbooks in the hands of students needs elaboration, summation, and alternate presention by the teacher. Many high school students do not have the patience and discipline to study carefully textbook material with the view of mastering the subject on their own. The lessons outlined in the following chapters are my attempts at such summation, elaboration, and alternate presentation.

Every so often we find a concept that is particularly difficult for most students to digest. It then becomes necessary to devise some method of helping them over these rough spots. I have tried in this book to point out these places and to indicate what I usually do to help them see the light. Sometimes it is necessary, when one knows there is a difficulty ahead, to emphasize a point that may be trivial in the context of the lesson then being taught in order to use this emphasis to help clear the road ahead when one arrives at the impasse. This, I have endeavored to do where my experience indicates its necessity.

While the book is aimed primarily at the high school physics teacher, we hope the other members of the science department will examine it. The chemistry teacher may find the sections on quantum and spectroscopy, for instance, useful while the teacher of physical science at the ninth grade level may be able to use some of these lessons a bit more qualitatively, for presentation to his students.

An attempt has been made, wherever possible, to suggest some alternate methods of presenting material. This will probably prove useful to both teachers and students.

I still meet many former institute participants at meetings of our local section of the American Association of Physics Teachers, and much of our conversation still revolves about different ways of presenting an idea so that it becomes meaningful to more students. We are all concerned about the numbers of intellectually capable students who do not elect to take physics, and we fondly hope that if we could make our course more exciting and meaningful, maybe the word will get out and more students will elect physics.

Part of our discussions also hinge on how we give our students grades, or the relative value of PSSC physics versus a more conventional teaching approach. One chapter has been devoted to a summary of these many conversations and some of the conclusions most of us agree on.

David Kutliroff

Contents

1

Teaching Physics—A Point of View

The truth you teach is not something that you have just discovered, and yet it is a very great contribution, I believe, to the vigor and beauty of the society in which we live.

ROBERT OPPENHEIMER

MAN'S HISTORY ON EARTH is marked with his quest for knowledge. His methods were often crude and his conclusions were often wrong. As his understanding of nature increased, however, his methods for wresting her secrets improved so that as he gained insight he discovered that there was always more and more to learn and found himself learning faster and faster.

Today the storehouse of scientific knowledge is so great and our technology has so many branches, that it is impossible to digest it all in a lifetime, let alone the short time students will spend in school. It is, therefore, necessary in the time they spend in formal study, to equip them with a sound foundation in basic concepts upon which to build the knowledge obtained by experience and further independent study and to train them to reason logically and objectively, so that they can arrive at independent conclusions based on known facts and new experimental evidence.

Students, therefore, should not be made to feel that they have to memorize pages of information. They should, however, try to follow your reasoning and be encouraged to draw their own conclusions as you lead them along toward the concepts to which our scientific signposts point. If they feel like arguing with you as you go along, please make them feel free to do so. Be thorough and rigorous in your explanations, and if you find yourself in unknown territory admit you need time to think and to research.

Most of the time you will get more out of it than your students will, but that is part of the fun that goes along with teaching physics.

I can recall several times when I went home trying to work out in my mind the answer to a puzzle broached to me by a student who had presented something in a way I had never thought of before. Sometimes I would stay awake half the night thinking about it when a "brilliant" solution would become apparent and I could hardly wait until the third period the next day to tell Johnny how to resolve his problem. Finally, both the third period and Johnny arrived, and I said, "Johnny, about that question you asked yesterday" and Johnny replied, "What question?"

Use mathematical symbolism where it becomes easier to make yourself understood in this manner. If it is properly introduced, students will learn not to be frightened by such symbolism and will learn to use it with enjoyment as a means of communication.

We report, in short, that our objective is to present a physics course designed to encourage logical deduction from empirical evidence, and to emphasize concept and understanding. Our aim would be to instill in our students an intellectual curiousity about natural phenomena; an ability to think logically; and an understanding of how a scientist arrives at his conclusions, even though they, as non-scientists, may not have been able to arrive at these conclusions themselves.

We know that it is important for all college-bound students, with intentions of continuing in any branch of science, to take physics, the most basic of sciences. Any intellectually capable high school student also should be strongly urged to elect physics, since this may be his last formal, academic exposure to a science course. To be a truly educated person in today's society, one must have an appreciation and understanding of science and the rigorous thinking a scientist must be trained to do. An appreciation of deductive reasoning should also be part of the background of every mature citizen.

Choices and Decisions

Innovations in physics teaching have done much to help teachers emphasize the approach we are espousing.

The first current innovation was the course designed by the Physical Science Study Committee. Its impact has been felt not only in physics teaching but has influenced and encouraged changes in the teaching of other sciences, math, and the humanities. Physics, prior to the introduction of the PSSC course, was usually taught as a series of disconnected topics in random order. Much emphasis was placed on technology and application. The contribution of the PSSC course which helped it gain worldwide acceptance and influenced the revision of other standard texts was the presentation of physics as a unified science. Students were no longer required to put the jig-saw puzzle together themselves. The emphasis on basic principles with application, brought in only when necessary to illustrate a point, made the course a rich experience for students and helped most teachers who were probably already looking for such an approach.

The course was designed as a joint venture by university and high school physics teachers. A new text was written, and a laboratory program was designed to go along with this approach. The cook book approach to laboratory was strictly avoided. Lab instructions were minimal and much of the help given students was carried by leading questions. The original intention was to have students come into the laboratory approaching each experiment like a research project. Most teachers, after some experience with the course, felt that most students need more than is offered in the laboratory manual and by selective feeding of required information or by asking leading questions all students can be made to complete their laboratory assignments with some profit to themselves.

Remember that any course you teach is taught at the level of which you and your students are capable. We cannot doubt that a physics teacher with a strong background and a healthy respect for his subject matter probably teaches physics with as much rigor and depth using conventional materials as he would teaching the PSSC course. It might be suggested, however, that though a student earning a D in PSSC physics might also have earned a D in a conventional course, he actually may have learned a little more physics because of the type of emphasis placed by the course material. Most would also agree that a great deal of the enthusiasm engendered in students is largely due to the enthusiasm the teacher shows in his presentations. It is, therefore, important that various approaches to teaching physics be available and acceptable, but bear in mind that this is as important for teachers as it is for students. Any approach that is used must be brought to your student with excitement and enthusiasm. The teacher must like what he is doing, and one way to get to like what you are doing is to become very familiar with your subject matter. No one can enjoy doing something that makes him uncomfortable.

Several sets of objective tests have been developed to go along with the course material. These tests are designed to emphasize the importance of concept and are most frustrating to the student who depends entirely on memorization of facts for performance on an achievement test. The tests are most useful as teaching devices if the teacher will go over them item by item the very next day showing how the correct answers could be obtained. After "sweating" over these same questions the day before, students are generally much more receptive than they are when discussing course material in the usual classroom situation.

A strong point of the PSSC textbook is the large number of good homework questions at the end of each chapter. The teacher's guide will help the new teacher select the most appropriate problems with various degrees of difficulty. It is recommended that three or four easy problems be assigned to help build confidence, some problems of minimum hardness with which some of your students may need a little help, and one or two hard problems to challenge the most talented of your students. It might even be necessary to point out that problems which have been assigned and gone over by students, though not successfully, lead to a much more meaningful learning situation when these problems are used to discuss the principles involved, than when you just lecture on an area which they had not yet had a chance to partially assimilate themselves.

Finally, it is generally known that the PSSC course is aimed at the student who normally takes physics in high school at the college preparatory level. Many teachers, however, have reported that they have been successful in using the suggested approach and even some of the materials for terminal students with good results. This probably reinforces one of our earlier statements regarding the effect of the teacher's enthusiasm and knowledge of his subject on his students.

Some concern with the apparent dropping enrollment of high school students in physics has led to the development of other physics courses which the authors feel might be palatable to those students who would not usually elect the physics course offered in their school.

One of the foremost of these approaches is what was formerly called Harvard Project Physics (now officially designated Project Physics). This course is also designed primarily for the college preparatory high school student but hopefully will appeal to more of those who are interested in humanities and, therefore, might not elect PSSC physics because of its lack of emphasis in this area. The course, designed by the Project Physics group, is essentially an introductory course in physics with some emphasis on the historical and philosophical aspects of the science. It was felt that it is not enough to study a science for its own sake, it is also important to see how man's progress in knowledge has affected the structure of his society, and some effort is made to show how physics interwines with and is necessary for complete understanding of other fields of human endeavor. Whether or not you are in complete sympathy with this approach, it is certain that there is much that a good teacher might choose to use to enrich whatever physics course he elects to teach.

Among the contributions made by the developers of the Project Physics course is the extensive use of new teaching aids like the "single concept" film loop. Use of these loops, both as part of a teacher's lecture and demonstration and independently by the student for self-study and in a laboratory exercise, is well worth investigating. It is a versatile tool which may well prove to be a revolutionary teaching technique for some teachers. Many loops have been developed by Project Physics and more will doubtlessly be produced by such groups and by commercial producers in the field of educational films.

The use of transparencies for overhead projection is also included as a recommended technique. These transparencies have been made available by Project Physics and are available also for PSSC and many of the topics taught in some of the older, more conventional approaches. Many teachers may prefer to make their own transparencies.

A third current approach to teaching high school physics is that devised by the Engineering Concept Curriculum Project. Strictly speaking, this was not intended by the authors to be competitive with the PSSC and HPP courses referred to above. Here, too, the authors were concerned about a lagging enrollment in physics. The thought occurs that the engineering viewpoint had been neglected. The PSSC emphasizes basic concepts in physics while the HPP, which is still concerned with most of the same basic concepts, points up a more humanistic approach. The authors of Engineering Concept Curriculum Project took the other extreme and assumed there was still a large number of students to whom an approach emphasizing the "man-made world" with its technology, and emphasizing man's attempts to mold his environment, might be more appealing. This course, it is stated by the authors, is in no way supposed to replace the regular physics course offered in a high school. It is to be offered as a supplementary course for twelfth graders who have already had a year of physics in the eleventh grade or as a different kind of course for those who would never have taken a physics course anyway.

It should be emphasized that a teacher should make it his business to familiarize himself thoroughly with all the approaches to teaching currently offered so that he can present a course most adaptable to the needs of his students and know it well enough so that he can teach it with confidence and enthusiasm. Possibly, he may select the materials of one as the course to offer his students, or he might use one of the courses as his basic approach and interlace it and enrich it with selected topics and presentations from the other courses. This must be done very carefully since the presentation of physics as a continuous whole must not be sacrificed for the sake of producing a hodge-podge of supposedly interesting topics. This can lead only to confusion.

Again, let me caution you, when you finally choose the course you wish to present, you will choose that which you think you can present most effectively and interestingly. Your students cannot possibly judge the difference between any two courses; they can only judge the difference between your presentations of two alternate courses.

Evaluation of Students

Much of student bias toward physics comes from successes their predecessors, including their older brothers and sisters and perhaps even their parents, have had in assimilating what was taught and in earning a grade which was compatible with their standards of success.

Your marking procedure might be one of the things which ought to be examined. Are you realistic? Does the mark you give a student actually mirror his accomplishments? Are your students penalized for taking physics because you are setting standards beyond the reach of most of them? A copy of my grading procedure which I give my students at the beginning of the year follows. Hopefully, it lets them know where I stand and what I expect of them. It works fairly well and I have found, by checking on grades in physics my graduates get when they continue in physics at those colleges who report back to us, that the marks they received were generally indicative and predicted their performance in subsequent courses.

HOW ONE ACQUIRES GRADES IN PHYSICS AND WHAT THESE GRADES MEAN

E. This is the easiest grade to get—one need only be on the rolls. You never volunteer, hardly ever turn in homework or lab reports and show by your performance on tests that you have successfully avoided opening the textbook.

D. Grades of D can be achieved in two ways—

1. Do well on exams and tests but never, or at least very rarely, turn in homework and/or laboratory reports.

2. Do not pass tests and quizzes but indicate you are trying by at least turning in all assigned work as required.

A "D" indicates a pass in physics and will not prevent you from graduating but indicates that you did not perform on a college preparatory level.

C. A "C" indicates satisfactory college preparatory performance but suggests that you do not plan on a career in physics or engineering.
B and B+. A "B" performance in physics indicates good college preparatory level achievement and suggests that you may continue in the field of physics or a career in engineering if you are sufficiently interested to work hard.
A and A+. An "A" in physics is good work and might indicate a talent in this area. Continued work in physics or a physical science based discipline is recommended for you.

Fifty per cent of your raw score is arrived at by averaging homework and lab assignments.

The homework obligation is the easiest to comply with. The grade for homework is calculated simply by taking the number of assignments turned in, divided by the number assigned. Individual papers are not graded. They are only examined to see if you made an honest attempt at solving problems and to help me ascertain what type of problems you are having the most trouble with.

Laboratory reports are graded on the basis of ten points for a perfect paper. Usually, the marks range from 7–10 with the bulk falling around 8–8.5.

The other half of the raw score is derived from the average of quizzes and tests. Since your performance on the standardized examinations used at the completion of each section is reported as the number correct out of a possible 35, you might be interested in knowing how a grade might be arrived at with this information. If you consider each correct answer to be worth 1/35 of 40 possible points, and then add 60 to the result, you will arrive at an approximate evaluation whereby 10–13 correct answers are equivalent to a D grade, 14–18 is a C grade, 19–22 might be necessary for a B, 23–26 for a B+, 27–31 for an A and 32–35 for an A+.

The total raw score is then the average of all the above. How then is the actual grade arrived at? This may be the same as the raw score or may be one grade above or below. This is where my unbiased professional judgment comes in. Did you turn in the required book report and was it good? Do you ask questions in class and do you sometimes contribute an intelligent and interesting argument? Do you actually work in the laboratory or do you just stand around taking notes?

Good luck—*D. Kutliroff*

Most high school guidance departments receive reports of first semester grades from the colleges at which their graduates are enrolled. If such an arrangement has not been made, it should be sought. The high school physics teachers should keep track of these grades and compare them with the mark they gave these students. Those students who elect to major in physics or engineering in college generally take physics in their freshman year. If your marks were realistic, "C" students would show a high mortality in a rigorous physics course designed for physics and math majors and future engineers. "B" students would show a high degree of success with less than 20 per cent failure. "A" students would prove your confidence in them by almost invariant success in introductory college physics courses.

Setting the Stage

Lastly, in an introductory chapter like this, the correct atmosphere should be set at once.

A short story I use when meeting my students for the first time illustrates, I think, an attitude which must be fostered and encouraged for the rest of the year.

We look at a leaf falling from a tree and see how long it takes to reach the ground. We drop a rock and watch it plummet down. We might be encouraged to follow the "common sense" approach and deduce that objects as light as leaves fall slowly and objects as heavy as rocks fall rapidly. Therefore, the heavier an object is, the faster it will fall. Galileo may have argued this topic with one of the Aristotelian church officials of his time.

"If I drop a small musket ball and a large cannon ball from a high tower," asks Galileo, "which will hit the ground first?"

"The cannon ball, of course," exclaims the eminent one, proud of his common sense.

"Now, if I tied the little musket ball and the very large cannon ball to opposite ends of a long rope and dropped them together, would they fall as fast as the large cannon ball falling by itself, as slow as the small musket ball falling alone or somewhere in between since the faster falling cannon ball is being slowed down by dragging the little musket ball after it?"

"Hmm," ponders the cleric, not yet seeing the trap the wily Galileo was weaving for him, "it will probably be the last case and fall at a rate somewhere between that of the cannon ball alone and the musket ball by itself."

"Now," says Galileo, "suppose I now tied the musket ball and cannon ball together in one tight bundle. Would it fall faster than the cannon ball by itself because it is now a heavier package or would it fall slower because it is still a large cannon ball dragging a slower moving little musket ball?"

The more familiar version of the story which exemplifies the experimental approach has Galileo dropping these weights from the tower of Pisa.

At least it is important to establish early that "common sense" is not always a reliable criterion as to what is right and wrong. We cannot argue against experiment or empirical proof. We can only try to explain evidence.

Let us define "common sense." Common sense is the effect of the sum of experience, environment, and lifetime contacts. What one person calls "common sense" is not necessarily the same to another with a dissimilar background. To the primitive Norseman it was "common sense" that a lightning bolt was the result of an activity of one of the gods in his mythology.

A useful demonstration which emphasizes the fallacy of the "common sense" approach and urges faith in physical principles might be prepared as follows:

Make an Indian Fakir board by putting tenpenny nails in a one square foot board approximately one centimeter apart. An easy way to prepare this board is to cement a piece of rectilinear graph paper to a piece of wood and with an electric drill make holes in rows and columns each about one centimeter from the other using the lines on the graph paper as a guide. When the holes have all been drilled (just large enough to accept the tenpenny nails with tight friction fit), put a nail in each hole. It should be cautioned here that the nails should be as uniform as possible. Discard any nails which are longer or shorter than the normal tenpenny nail.

With the board now on display in front of the class, one can point out that it is common knowledge that it is painful to sit on a nail. According to the "common sense" approach of some, to sit on two nails ought to hurt twice as much. You can then explain that according to physical principles you know that you can sit on these nails comfortably, and as you sit on the points of the nails discuss the need for faith—faith in physical principles. Now, as you remain seated comfortably on your seat of nails, explain that pain is caused by excessive pressure and that pressure is defined a force per unit area. The force which is your weight when concentrated on an area as small as the point of a nail will indeed be excessive but when distributed over a large number of nail points would make for a very small force per unit area.

If you wish, you may even stand on this board in bare feet without too much discomfort and then invite your students to demonstrate their faith in physics by repeating your performance.

In order to communicate ideas, reasoning should be based directly on experimental evidence. Thus, we will help our students build a new "common sense" and perhaps learn to reason in all fields of human activity with a minimum of bias or perconceived notions. Knowing that it is a human failing to be biased, that "common sense" is not always trustworthy, perhaps we can help develop a citizenry who will be less dogmatic in attitude and more willing to listen to the other side of an argument.

The Model

Some evidence must be of an indirect nature. The concept of a model must be used and understood. How do we foretell the existence of the tiny particles of which matter is made up? How can we talk about the distance or size of stars which are so far away that it takes millions of years for their light to reach us?

We build mental models of the universe in areas too small and too large to measure directly. We teach our students to accept these models as long as they fit the observed facts and to use these models to predict further phenomena. They must then accept our models, but always with the reservation that a new discovery may make it necessary to change our model.

A pertinent observation by Albert Einstein points up this attitude. He stated that no amount of experiments could ever prove his theory, but it would take only one experiment to disprove it.

It is also useful to point out that nature and its laws are observed in a laboratory, but a laboratory may be a room containing special tables, elaborate instruments and measuring devices, or it may be a field, a forest or anywhere else physical measurements are made. Simply, a physics laboratory must be defined as a place where we make measurements for further refinement of our knowledge of the physical world.

2

A Laboratory Is for Making Measurements

I often say that when you can measure what you are speaking about, and express it in numbers, you know something about it; but when you cannot express it in numbers, your knowledge is of a meagre and unsatisfactory kind; it may be the beginning of knowledge, but you have scarcely, in your thoughts, advanced to the stage of Science, whatever the matter may be.

LORD KELVIN

IT IS IMPORTANT TO EMPHASIZE from the start that physics is a quantitative science. The physics laboratory is a place to make measurements. These measurements are then used to look for relationships which will help us predict, by extrapolating or interpolating, how dependent variables are affected when one or more parameters are changed.

Since, in physics, we deal with measurements ranging from distances on an interstellar scale to the spaces between subatomic particles, it is necessary to accustom our students to the use of exponential notation in the handling of numbers. Without this, many calculations and interpretations become unnecessary drudgery. Most measurements must be arrived at indirectly. It must also be emphasized that in reporting measurements and the conclusions reached from these measurements, the only data to be reported must be within the precision obtainable in the measurements. The subject of significant figures will be discussed shortly in more detail.

The necessity to base much of our understanding of the physical sciences on indirect measurements justifies spending some time at the beginning of an introductory course in physics familiarizing students with some of the techniques.

Following are some simple laboratory exercises which might be useful.

Triangulations

Students are brought out to the school parking lot or athletic field and asked to measure the distance between two distant objects without actually laying out a tape between these two objects. (The two objects selected may be two trees or a telephone pole and a car, etc.) Going outside for an early laboratory emphasizes the idea that a laboratory is not necessarily a room filled with stainless steel instruments. It can be any place where measurements can be made.

A base line will be selected and measurements will be made by triangulation. The only equipment required is a cork board, or soft wood board which can hold straight pins, paper and pencil, and a meter stick. Some laboratory exercises employ the use of simple range finders, parallax viewers, etc., but it would be well to emphasize that these are just methods of triangulation with limited base lines. The emphasis on such an introductory laboratory excercise should be on having the student understand what he is doing rather than rely on working with what is to him at this stage, a "black box" instrument.

Two students usually work together on this problem. Their cork board with a paper tacked on it must always have one side parallel to the base line and a corresponding base line drawn on the paper parallel to that side. One student then stands on a spot on the base line which he marks and calls A. He marks a similar point on the base line on the paper and labels it A^1. He then sights with straight pins to the two distant objects thus constructing the viewing angles at one end of his base line between the base line and the lines of sight to the two distant objects. (See Figure 2–1.)

The partner's function at this point is to make sure that the base line on the sighting tablet is always parallel to the actual base line while the first student is constructing the angles. They then walk over to the other end of their selected base line which might be called B and mark the point B^1 on the corresponding base line on the sighting pad. The sight lines are drawn and the sighting angles are constructed similar to the way the sight lines were found from point A. The sight line from both ends of the base line are then extended until they intersect. The intersection of the sight lines on the paper gives the positions of the two objects sighted in real space.

Since simple geometry can indicate that all the figures on the paper are similar to the geometry of the actual spaces measured, and corresponding

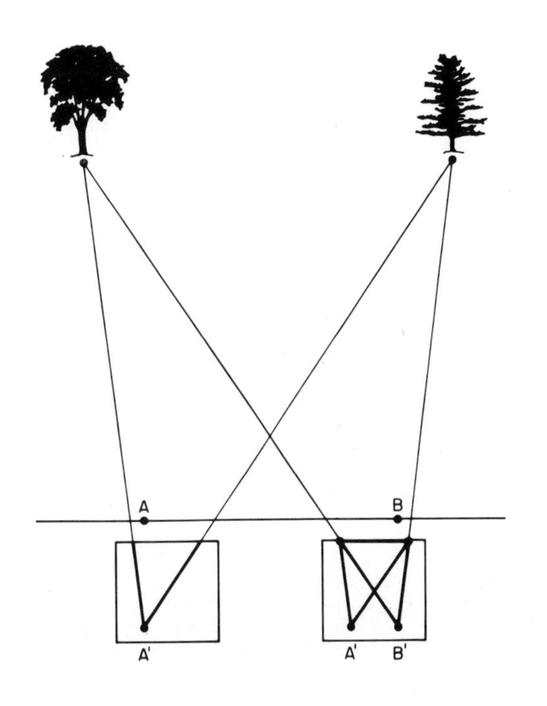

Figure 2-1

parts of similar figures have the same ratio, the student can, by measuring the base lines A^1B^1 on the paper and the actual base line AB from whose ends he made the sightings, determine the distance between the two distant objects to which he sighted or the distance from either of these objects to any point on the base line. After completing his figure, the student ought to be encouraged to make the direct measurement to check himself and calculate the per cent error.

Parallax

A method of indirect measurement utilizing the parallax technique for measuring the distance to nearby stars can be similarly performed. The student will be asked to sight from one of his base lines to a relatively nearby object which is lined up with a far distant object. He lines up two pins on his sighting board to indicate the line of sight from this point. Moving over to the other end of his chosen base line, he makes sure that his line of sight is still maintained with the far distant (fixed) object. He places another pin on the base line on his paper so that it lines up with his head pin and the nearby object. (See Figure 2–2.)

The similarity of triangles can be readily seen and the distances calculated. The similarity between this procedure and the method of measuring the distance to nearby stars, using the diameter of the earth's orbit as a

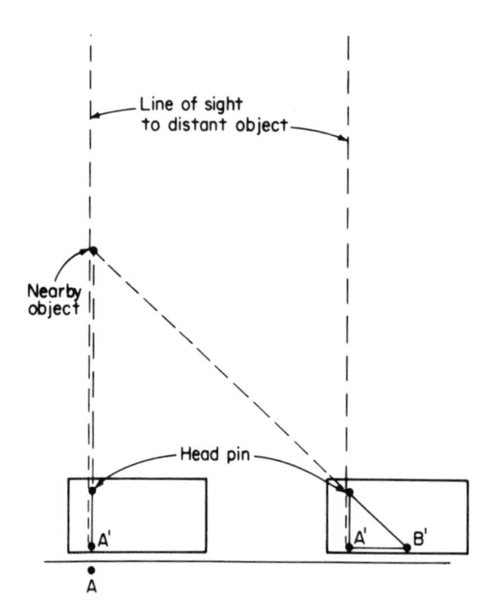

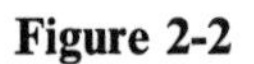

Figure 2-2

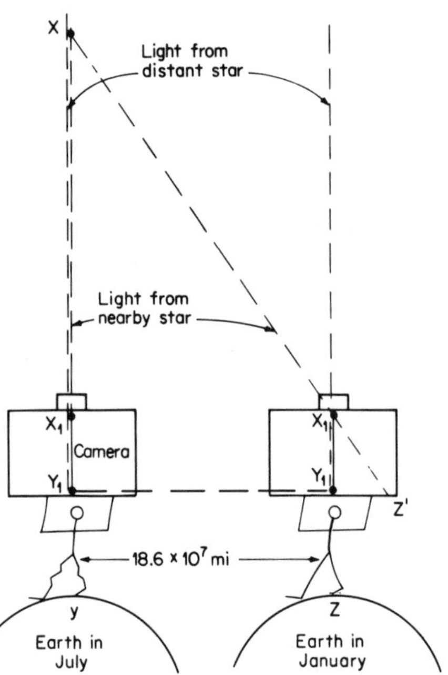

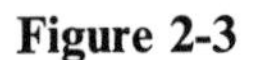

Figure 2-3

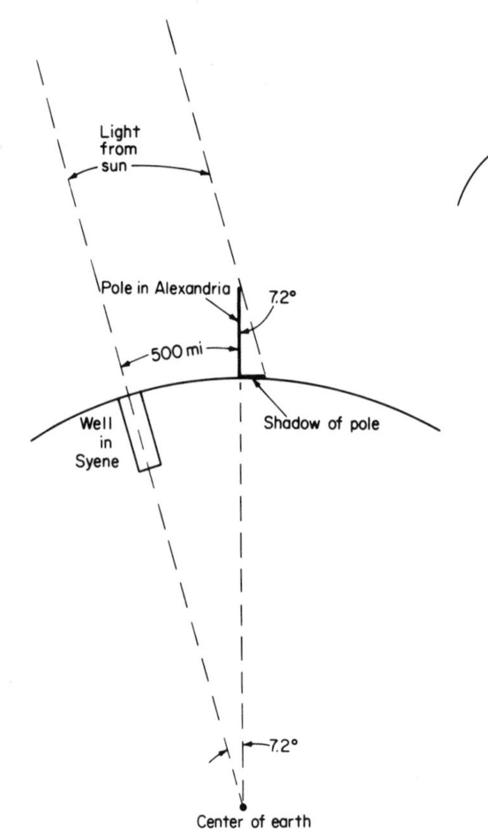

Figure 2-4

base line and measuring the parallax shift between the nearby star and the "fixed stars," should be pointed out. A sample lesson illustrating such use of parallax in astronomical measurement follows:

One method of indirect measurement makes use of a phenomenon called parallax. Your students have all experienced parallax without giving it a name. For instance, if one is in a moving train and gazes out the window, it can be noticed that the nearby telephone poles appear to be whizzing by, distant houses appear to be moving much more slowly with respect to the observer, while perhaps a distant mountain doesn't appear to change its position at all even though one knows that the train must be moving about 60 miles per hour. The further an object is, the less one's angle of vision changes as one changes his position relative to it.

Suppose, therefore, one wished to measure the distance of a star. Some stars are so far away from our solar system that they always appear to be in the same position relative to each other or to subtend the same angle of vision even though the earth sweeps through an orbit around the sun of about 93×10^6 miles in radius. These are called fixed stars. Other stars, relatively much closer, will appear to move relative to these. (See Figure 2–3.)

Pictures are taken of a section of the sky at six month intervals so that the distance between the base line positions for taking the picture is 186×10^6 miles (approximately the diameter of the earth's orbit). ZY equals 186×10^6 miles, x^1y^1 is the distance of the camera lens to the photographic plate in the camera, y^1z^1 is the measured change in relative position of the two stars on the photographic plate, and yx is the distance of the nearby star from the earth. Since triangles zyx and $z^1y^1x^1$ are similar, $yx = yz \cdot x^1y^1/y^1z^1$. With the quantity yx being the only unknown, the solution of the problem is relatively simple.

A classic example of ingenuity in measurements is the method used by Eratosthenes of Alexandria to measure the circumference of the earth. It makes an interesting illustration for students. (See Figure 2–4.)

He noted that, in the city of Syene which is on the Tropic of Cancer, the sun shown directly down deep wells at noon on July 21, while at the same time about 500 miles due north in the city of Alexandria, a pole standing perpendicular to the earth's surface cast a shadow.

By measuring the height of the pole and the length of the shadow, he knew the tangent of the angle by which the sun's rays were not parallel to the pole in Alexandria. Therefore, he knew the angle made between extensions of the pole and the extension of the line of the well in Syene to the center of the earth, where these two lines intersected. This angle turned out to be 7.2 degrees or about 1/50 of 360°.

Five hundred miles is, therefore, 1/50 the circumference of the earth (if we assume a spherical shape) and the circumference of the earth is, therefore, approximately $50 \times 500 = 25{,}000$ miles.

Irregular Areas

Another exercise in indirect measurements involves the measurement of irregular areas. Students might be asked to find the area of the county in which they live and to use at least two different methods of computing this area. It might be assigned as a home laboratory problem. Having your students go through a laboratory assignment at home once in a while is a good practice in that it again emphasizes that the location of the laboratory is unimportant as long as it is a convenient place to make measurements.

A convenient method of finding such an area might be to obtain a road map and cut out the county whose area is to be determined. Trace the outline of the county on a piece of graph paper and determine the area represented by each square by the scale given on the map. The area of the county can then be determined by counting squares and estimating partial squares.

Another method of determining the area would be by weighing it. Students can be shown how to make a sensitive equal arm balance from a soda straw, a straight pin, some razor blades, and a little aluminum foil. (See Figure 2–5.)

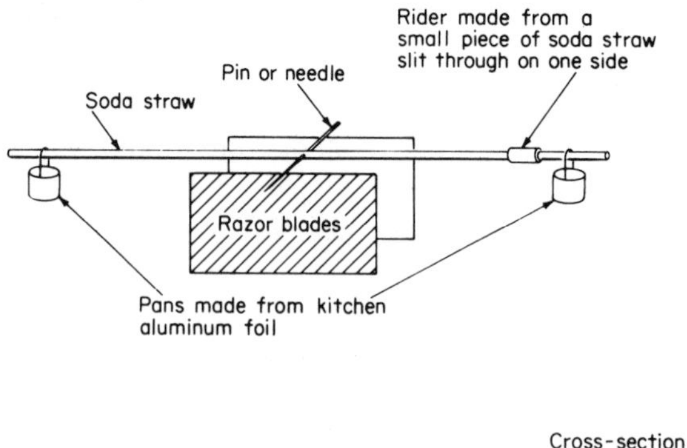

Cross-section of soda straw (note it is pierced above center of mass)
Needle
Wooden block with razor blade supported by rubber band

Fig. 2.5

The balance once recommended in the very early versions of the PSSC laboratory manual would be quite satisfactory. One assumes, of course, that the road map is made of a fairly homogenous paper of common thickness and density. The cut-out section is then placed on a pan of the homemade equal arm balance and cut-out squares of the graph paper are then placed in the other pan until they balance. A square of road map paper whose area is known because of the marked scale of the map is then similarly weighed and the area of the irregular surface can be determined because the ratio of the two areas is equal to the ratio of their weights. Incidentally, it would be useful to point out that the units of weight are completely arbitrary. In this case the areas are weighed in units of graph paper squares.

An extension of these exercises in irregular area determination that has often proved interesting to students is the determination of the value of π.

A circle of radius r is cut out of a piece of paper and a square of side length $= r$ is also cut out. The relative areas of the circle and the square are then determined by the weighing method outlined above or the method of counting squares previously discussed. I hope one is not too surprised to find that the area of the circle is about 3.14 times the area of the square.

Measuring Very Small Distances

Finding the thickness of a page in a book or the thickness of a razor blade by measuring the thickness of a stack and dividing by the number in the stack is another excellent example of indirect measurement measurement.

A very good exercise in indirect measurement of a very small distance is the estimation of the length of a molecule by measuring the thickness of a monolayer. It might be interesting to outline the usual procedure and then suggest some student extensions to this exercise.

Before assigning this laboratory, one ought to explain to students why physical chemists predict that fatty acids like oleic and stearic will form monolayers. If they have had chemistry before taking physics, one can show that in the structure $R - C^{=O}_{-OH}$, the organic acid is water soluble while the long carbon chain is not, so that one end of the molecule may be considered watertropic while the other end may be thought to be waterphobic. An analogy may be made to feeding pigs at a long trough where one end of the pig likes food while the other end does not. Therefore, pigs fed at a long trough form a monoporcine layer as long as there is room enough for them all to get to the trough.

A pair of students in each section can make up the stock solution. First put 5cc of oleic acid in 95cc ethyl alcohol, thus making up 100cc of 5 per cent oleic acid in alcohol. Now take 5cc of this solution and put it into 45cc of alcohol, thus making 10 per cent of a 5 per cent solution of oleic acid in alcohol. The stock solution is now put into dropper bottles and each student pair can measure the size of a drop by seeing how many drops will make up a 1cc volume in a small graduate. It will be found that a drop is about 1/50cc in volume. A drop of this solution is then squeezed out on the surface of water in a fairly large tray and the diameter of the slick is measured. Since the slick formed will be approximately 20cm in diameter, the tray ought to be large enough to hold it.

Ripple tanks make fine trays for the experiment. The only precautions that need be pointed out here are that when they are emptied, the water should be poured out over the side rather than drained through the side hole with which some ripple tanks are provided. This is to prevent an oily deposit being left on the glass.

To see the slick, one usually dusts the surface of the water with lycopedium powder. Chalk dust dusted from an eraser is just as satisfactory. The only caution one must give students is to use the dust sparingly. The tendency is to dust too liberally and this prevents circular smoothly edged slicks from being formed. If the slick is not a perfect circle, then one measures several diameters and takes an average. With a slick radius of 10cm, the calculations might look like the following:

$2 \times 10^{-2}\text{cm}^3/\text{drop} \times 5 \times 10^{-3}$ oleic acid $= 10 \times 10^{-5}\text{cm}^3$ oleic acid/drop.

$\pi r^2 = 3 \times 10^2\,\text{cm}^2$ (area of slick)

$10 \times 10^{-5}\text{cm}^3/3 \times 10^2\text{cm}^2 \cong 3 \times 10^{-7}\text{cm}$

It might be pointed out that 10^{-7}cm is about the right order of magnitude for a large molecule. An atomic diameter would be in the order of one angstrom or 10^{-8}cm. 10^{-7}cm would indicate a chain of approximately 10 atoms. The interesting thing is that the student should know this and be able to see if his experimental results are reasonable.

It can be further pointed out that if you have helped him visualize the molecule as essentially a string of carbon atoms and have informed him that the atomic mass of oleic acid is 282, then he has enough information to predict the approximate size of the atom before confirming it experimentally. Knowing the carbon atom has a mass of 12 a m u's, then $282/12 \approx 20^{+}$. Since atoms are roughly one angstrom in diameter, then 20 A $\approx$ approximately 2×10^{-7}cm. Such reasoning should be encouraged. Certainly one should not ignore what is already known before proceeding with an experiment. If experimental results do not confirm the predicted results based on the accepted model, then one ought to reexamine the model (assuming our experimental techniques were faultless).

A suggested extension is to estimate the oleic acid molecule's shape to be a cube and, therefore a volume of $27 \times 10^{-21}\text{cm}^3$.

If oleic acid has a density of .89 gm/cm^3 and a mole has a mass of 282 gms, then how many molecules would there be in a mole?

$$\frac{282 \text{ gms/mole}}{.89 \text{ gm/cm}^3} = 320 \text{ cm}^3\text{/mole}$$

then $$\frac{320 \text{ cm}^3\text{/mole}}{27 \times 10^{-21} \text{ cm}^3\text{/molecule}} = 1.2 \times 10^{22} \text{ molecules/moles}$$

This is off by a little over an order of magnitude, the error being in the original estimation of the shape of the molecule. It is, however, a step toward experimental verification of Avogadro's Number.

Another interesting problem would be to suggest the shape of the molecule, assuming you know Avogadro's Number. If the volume of the slick containing a whole mole of molecules is 320 cm^3 as previously calculated, then $320 \text{ cm}^3/3 \times 10^{-7} \text{ cm} = 1.07 \times 10^9 \text{cm}^2$ for the area of this slick. $10.7 \times 10^8/6 \times 10^{23} = 1.8 \times 10^{-15}$ or $18 \times 10^{-16} \text{ cm}^2$ for the cross-sectional area if each molecule. The diameter of these molecules then is shown as follows:

$$\pi R^2 = 18 \times 10^{-16} \text{ cm}^2$$
$$R^2 = 5.7 \times 10^{-16} \text{ cm}^2$$
$$R = 2.4 \times 10^{-8} \text{ cm}$$
$$D = 4.8 \times 10^{-8} \text{ cm}$$

The oleic acid molecule is approximately ten times longer than its thickness, and the fact that its thickness is in the order of magnitude of an angstrom would attest to the fact that it may very well be a single stranded chain of carbon atoms.

In order to interpret results of the laboratory exercises previously suggested, the use of exponential notation and the concepts of measurements must first be taught in the classroom. Following are some sample lessons which might prove useful.

Exponential Notation

Before your students actually begin to study physics, they will first have to learn a few things about making measurements and use them in drawing conclusions. Let them first consider the problem of handling very large and very small numbers. Consider that they may talk about distances to stars wherein the light takes a million or more years to get to us.

If light travels 300,000,000 meters per second and there are 3600 seconds in an hour, 24 hours in a day and 365 days in a year, going through the calculations to find the number of meters in one light year and just writing the final number with all its digits and zeros, would be laborious indeed. We have an easier way to do this. Instead of 300,000,000, we write 3.0×10^8. This means $3 \times$ (10 multiplied by itself eight times). 3600 can be written as 3.6×10^3. Now, instead of multiplying 300,000,000 by 3600, we can multiply 3×10^8 by 3.6×10^3 or $3.0 \times 3.6 \times 10^8 \times 10^3$. $3.0 \times 3.6 = 10.8$ and $10^8 \times 10^3 = 10^{11}$ or 10.8×10^{11} (10 multiplied by itself 8 times, times 10 multiplied by itself 3 times equals 10 multiplied by itself eleven times).

In multiplying powers of ten, we merely add the exponents. Let us then go through the entire calculation.

$$3.0 \times 10^8 \times 3.6 \times 10^3 \times 2.4 \times 10^1 \times 3.7 \times 10^3 = 9560 \times 10^{15}$$
$$= 96 \times 10^2 \times 10^{15} = 96 \times 10^{17} = 9.6 \times 10^{18} \text{ meters.}$$

Think of writing 96 with 17 zeros following it.

By the same token, we may have to consider extremely small distances. For instance, the average distance between the molecules of air in your room is $1/10^8$ meters. Let us put down a few equivalents so that you can see how the notation works.

$$\frac{1}{10} = \frac{1}{10^1} = .1 = 10^{-1}$$
$$\frac{1}{100} = \frac{1}{10^2} = .01 = 10^{-2}$$
$$\frac{1}{1000} = \frac{1}{10^3} = .001 = 10^{-3}$$

Now if we multiply $1/10 \times 1000$ we get 100. Similarly $10^{-1} \times 10^3 = 10^2$. We still are multiplying by adding exponents.

$$\begin{array}{r} -1 \\ +3 \\ \hline +2 \end{array}$$

Assign practice problems involving exponential notation. Your students will find the experience helpful. It is especially useful in slide rule calculations.

Units

Measurements in length, volume, time, and mass are all comparisons to some standard. An inch might have been the length of the last joint of an English king's thumb. A day is the length of time it takes for the earth to revolve once on its axis. A meter is what was once calculated to be 10^{-7} of the distance from the equator to the north pole. A gram is the mass of 1 cubic centimeter of water under certain duplicatable conditions. Whenever we make a measurement, we compare directly or indirectly with a standard. The precision of our results depends on the instrument used in the measurement and human fallibility. Every meaurement we make is an estimate made under these conditions.

There are various units of measurement which have evolved by necessity as men became conscious of the need for them. We are familiar with the English system of measurements with its irregularities, three feet in a yard, 12 inches per foot, volume measurements in quarts and gallons, or cubic feet, etc. Even when we use this system all the time, we must have memorized or know where to look up conversion tables in order to make necessary calculations.

The French people, after their revolution, were determined to start anew in all fields. One of their contributions was the metric system. This system of measurements did not just evolve. It was planned. You might compare the English system and metric system to an old city with its crooked streets, nonsystematic street names and numbers, a city where only an old inhabitant can find his way about, to a planned city with an orderly arrangement that any stranger can move around in with confidence once he has taken the trouble to learn the system.

Once students have learned it, they will find the metric system of measurement very easy to use. It is based on increments of ten. (Exponential notation will fit in very nicely in our calculations.)

Significant Numbers

Suppose you ask a student to measure the width of his desk with a meter stick. His measuring operation consists of comparing the width of his desk to the distance between two marks on his stick. The measuring stick is marked off in centimeters and millimeters. Perhaps he carefully lines up the one cm mark on the end of his ruler with one end of the desk. He then checks to see what mark lines up with the other end of the desk. He finds the desk is somewhere between 55.6 and 55.7 cm wide. He feels his best estimate is around 55.65 cm and reports it as such. He measures the length of the desk and finds it to be 137.85 cm and now he wishes to find the area thus represented. Before multiplying 55.65 by 137.85 cm, let him look at the significance of these figures. When he measured the width of his desk, he was sure of the 55.6; the last figure was an approximation. Ask him to circle all the approximate figures, perform his multiplication and see what happens.

```
       1 3 7 . 8⑤
         5 5 . 6⑤
      ───────────
       ⑥⑧⑨②⑤
     8 2 7 1⓪
   6 8 9 2⑤
 6 8 9 2⑤
 ──────────────
 7 6 7①③⑤②⑤
```

All the top row of figures are circled because they cannot be other than approximations since one of the multiplicants was an approximation. This is also true of the last figure in the second, third and fourth rows. The last five figures in the answer are circled as estimates since one or more of the numbers that entered into their summation was an estimated number. It would be wrong to report 7671.3525 sq cm. The correct answer is 7.671×10^3 cm^2. This answer, while not as precise as the previous one, is certainly more accurate because the first three figures are the only ones he is sure of and the fourth is an estimate.

Point out that the number of significant figures in the final product is the same as the number of significant figures in the smallest multiplicant. This is usually true and is a fairly safe rule to follow if you need a rule. However, it is easy to show exceptions to the rule. Emphasize the point that the number of significant figures you report are what you can justify and not because you are following a hard and fast rule.

For instance, suppose you are asked to report the area of a bench top 4.001 inches long and 91 inches wide. By your rule, the product ought to have only two significant figures. Let us see if we can justify this.

```
   4 0 0①
       9①
   ④⓪⓪①
 3 6 0 0⑨
 3 6④⓪⑨①
```

All we can say is that there are approximately 364,000 square inches of area to be reported. To emphasize the number of significant figures, we can report it as 36.4×10^4 or 3.64×10^5 sq cm. Note that there are three significant figures instead of two.

On the other hand, suppose you were asked to double the width of the desk. This means you would have to multiply 55.65 cm by 2. Does that mean that you would end up with only one significant number because of the two? No, the two is not a measured quantity. It is an arbitrary factor and is significant as 2,2.0,2.00,2.000 or as many places as you wish. Only the number of significant figures in the *measurement* governs the number of significant figures in the product. All this reasoning applies, of course, to dividing, adding, and subtracting. It is easier to make a general hard and fast rule for adding and subtracting. The number of significant figures to the right of the decimal point in the answer is that of the least amount to the right of the decimal in one of the numbers that contributed to the final answer.

Try this out with numbers you make up. Use the circle method we use above and remember that if an estimated number is part of a whole, then the whole must be considered an estimate.

In a discussion of significant figures, it ought to be pointed out that there are some measurements which can only be expressed as a power of ten. We call this an order of magnitude.

For instance, the distance to the North Star is about 10^{19} meters. To talk about it being 2×10^{19} or 8×10^{19} has no significance since we cannot measure to that accuracy. We must be satisfied with the knowledge that it is closer to 10^{19} meters than 10^{18} or 10^{20} meters. This order of magnitude measurement is the best information we have but it is still very important information.

On the macrocosmic scale we can talk about distances of 10^{22} meters to Andromeda, the nearest galaxy to our own, or 10^{20} meters, the distance of our sun from the center of our galaxy. In microcosmic measurements, we measure the dimensions of an atom in units of angstroms, an order of magnitude of 10^{-10} meters, the diameter of a proton is estimated at 10^{-15} meters, and the size of a molecule of oil is about 10^{-9} meters.

Finally, it should be pointed out that the use of exponential notation permits you to give more information than you could without it. For instance, it is generally reported that the distance from the earth to the sun is 93,000,000 miles. By reporting this as 9.3×10^7 or 9.30×10^7 or 9.300×10^7, we report not only the magnitude of the distance but the precision with which we know it.

3

Atomism and Chemistry

Little things have smaller things
That make them up inside them
And smaller things have lesser things
And so on ad infinitum

We emphasize the atomistic theory of matter and later the quantum nature of energy. Evidence indicates that there is no "ad infinitum" to which we can keep seeking the fundamental particle. The fundamental nature of the electron, proton, and other particles are recognized; the elementary charge and the photon of energy give further evidence of the graininess of nature as we begin to examine the microcosm.

It would be well to emphasize this idea from the beginning. A familiar example which might be cited is the seeming continuity of a sample of water which appears to be able to be subdivided indefinitely until you get down to a few molecules. You then must realize that no matter how much water you have it must be a whole number multiple of that fundamental quantum of water, the water molecule.

THE MOLE

Before continuing further, it might be useful to introduce the mole concept, not as some special and mysterious piece of technological jargon but merely as the name of a quantity of things.

The concept of Avogadro's number was first suggested in the proposal that any gas under the same conditions of temperature and pressure would contain the same number of particles. A gram molecular mass (or a gram atomic mass of a mono atomic gas) occupies 22.4 liters at 0° centigrade and sea level pressure of 14.7 pounds/sq.in. This quantity of gas has been defined as a mole and has been found experimentally to contain 6.025×10^{23} particles. It might be pointed out that Avogadro's hypothesis merely indicates that equal volumes of gas at the same temperatures and pressure contain equal numbers of molecules. Avogadro never knew Avogadro's number. The stress on the connection between 22.4 liters and the number 6.025×10^{23} is most unfortunate since this involves a very special case under very specific circumstances, yet many students come away with the feeling that this is a very important relationship.

It would be well merely to define a mole of anything as 6.025×10^{23} of these things. A mole of atoms is 6.025×10^{23} atoms and a mole of molecules is 6.025×10^{23} molecules. A mole of marbles is 6.025×10^{23} marbles.

It might also be useful to show your students that in problem solving, the number 6.025×10^{23} can be used as a proportionality constant to relate atomic mass units to grams in the same way that the number 454 relates grams to pounds. An analogy which illustrates this idea might be proposed as follows:

We know that there are 454 grams in a pound. Let us assume several different objects which we might call, for the sake of neutral illustration, a whatchit, a gizmo, and a whoozit. A gizmo has a mass of one gram. It occurs naturally in groups of two gizmos containing two grams of mass. A whatchit contains sixteen grams and occurs naturally in fundamental groups of two with thirty-two grams of mass in each unit. A whoozit is not a unit. It is a natural group of two gizmos and one whatchit containing eighteen grams. Similarily, there are $6,025 \times 10^{23}$ atomic mass units (AMU) in one gram.

Hydrogen atoms contain 1.008 AMUs each.
Hydrogen molecules (2 atoms) have a mass of 2.016 AMUs each.
Oxygen atoms are composed of 16 AMUs each.
Oxygen molecules (2 atoms) contain 32 AMUs in each unit of oxygen compound.
Water composed of two hydrogen atoms and one oxygen atom contains 18 AMUs each.

Below are two sets of statistics for easy comparison.

454 gizmo units would equal 1 lb. of gizmos
454 gizmo groups would equal 2 lbs. of gizmos

454 whatchit units would equal 16 lbs. of whatchits
454 whatchit groups would equal 32 lbs. of whatchits
454 whoozit groups would equal 18 lbs. of whoozits
6.025×10^{23} hydrogen atoms in 1.008 grams of hydrogen
6.025×10^{23} H_2 molecules in 2.016 grams of H_2
6.025×10^{23} oxygen atoms in 16 grams of oxygen
6.025×10^{23} O_2 molecules in 32 grams of O_2
6.025×10^{23} H_2O molecules in 18 grams of water

The concept of a mole being 6.025×10^{23} things should be no more complicated or harder to conceive of than the fact that a dozen is 12 things.

STATES OF MATTER

We know that matter exists in different states. Solids have stable sizes and shapes. Liquids have stable volumes, but with shapes determined by the container holding them. Gases exhibit complete fluidity and will take the size and shape of the container no matter how little is present and how large is the container. Finally, we have plasmas, in which the energy content is so high that they exist as a sort of gas of ions. This last, though least familiar to us, is probably the most common state of matter in the universe, since that is the state in which most matter exists in the stars, and that is where most of the matter in the universe is.

Let us explore these states of matter from an atomistic viewpoint. In a solid, the building blocks (atoms or molecules) are very close to each other. Since atoms attract each other, they will tend to remain rigidly in place. The attractive forces are strongest when the distance between the particles is smallest. As we add energy (heat) to the solid, the particles will vibrate, colliding with each other harder and harder, driving themselves apart from each other.

The forces holding atoms together in orderly agglomerates (crystals) become weaker. Breaking these bonds indicates a gain in potential energy and hence no rise in temperature. The individual atom becomes free to vibrate and move without having to drag several of its neighbors with it. The shape of the mass of atoms can now change and flow.

The substance is now in the liquid state, bumping each other and moving in random directions at varying unpredictable velocities. We cannot follow or account for the behavior of each individual particle; yet because there are so many millions of millions of particles present, their average behavior is always predictable, and we can express the energy level of the substance as its temperature. Since this is only an average, some individual particles will have enough energy to overcome the attractive forces which hold them in proximity to their neighbors, and will fly off into space, thus lowering the average energy content of the mass of material left behind. We know from experience how the temperature of a liquid is lowered by allowing it to evaporate. Let a student put some alcohol on the back of his hand and report the cooling affect as the alcohol evaporates.

As the addition of heat is continued, the particles in the liquid move faster and faster until they bump each other so far apart that the attraction between them is too weak to hold them together. As they move apart from each other gaining potential energy, there is no apparent temperature rise, indicating no gain in average kinetic energy. The point where this phase change occurs is manifested, therefore, by a temperature plateau, while heat is continuously being added. The constant temperature is called the boiling point of the liquid. The liquid now turns into a gas.

THE IDEAL GAS MODEL

Let us examine this gas. In order to keep all the gas for our scrutiny, the container must be closed. Any opening will allow these fast moving particles to escape. With the container closed, the gas has several physical characteristics which can be measured, namely its volume, temperature and pressure.

The volume is determined by the size of the vessel. The temperature is an expression of the average linear kinetic energy of the particles. This, in turn, is determined by the mass and the velocity at which these particles are moving about.

The pressure on the walls of the container is determined by the number of collisions per unit surface per unit time and the force exerted by each collision. Pressure is defined as force per unit area as in newtons per square meter, dynes per square centimeter or pounds per square inch. The force exerted by an individual particle on the wall depends on its momentum; that is, the product of its mass times its velocity. Therefore, a small mass having a high velocity can exert as much force as a large mass particle having a relatively small velocity. It is also obvious that particles having greater velocities will collide with the wall of the container more frequently.

Suppose we now examine the container of gas under our inspection and note what happens to the volume, temperature, or pressure as we change one or the other of these conditions.

Let us assume our container is a cylinder with a tightly fitting piston on one end. We can vary the volume by moving the piston. If we push the piston halfway in, assuming that the temperature remains constant, what happens to the pressure? We now have twice as many particles enclosed in the volume and the confining walls will receive twice as many collisions in the same time as they had previously. Our pressure is, therefore, doubled when the volume is cut in half. Similarly, if we expanded our container to double its volume, we would find that the pressure would be half its original volume.

In other words, we find that the pressure is inversely proportional to the volume, or the product of volume, and pressure is constant.

$$P_1/V_1 = P_2/V_2 = P_3/V_3 = K$$

Let us now vary our thought experiment. Suppose we keep the volume constant and change the temperature. Note that as we increase the temperature, the sides of the container are struck harder and more often. Careful experimental evidence indicates that the pressure varies directly with the absolute temperature. Thus, the ratio of absolute pressure to the absolute temperature remains constant if nothing else is varied.

$$\frac{P_1}{T_1} = \frac{P_2}{T_2} = \frac{P_3}{T_3} = K$$

It is also possible to show experimentally that if we keep the pressure constant, the volume will have to increase as we increase the temperature and decrease as we decrease the temperature.

In fact, we find that if we start at 0°C, the volume will be exactly doubled if we go up to 273°C. Similarly, if we start at 0°C and go down to −137°C, our volume would become half of what it was (if it hasn't become a liquid or solid before then). We can then extrapolate that the volume would be 0 when the temperature is 0° absolute (or Kelvin) or −273°C. (Of course, it would turn to a liquid or solid before then and not obey the gas laws.)

Our volume, therefore, varies directly with the temperature in degrees absolute.

$$\frac{V_1}{T_1} = \frac{V_2}{T_2} = \frac{V_3}{T_3} = K$$

We can combine these three relationships in one combined statement as follows:

$$\frac{P_1V_1}{T_1} = \frac{P_2V_2}{T_2} = \frac{P_3V_3}{T_3} = K$$

$$\text{or } PV = KT$$

$PV = KT$ is a simple statement of the Gas Laws. The constant K can be further broken down. For instance, we did not change the number of particles of the gas in the experiments described. Suppose we add gas in quantities of unit of moles. We can see that an added number of particles will increase the number of collisions, or the volume will have to be increased if the number of collisions per unit area is to be held constant. Similarly, if we want to hold the product of the pressure and volume constant, we will have to reduce the temperature. Experiments have been done to verify these relationships and, on a more sophisticated level, the Gas Law can be expressed as $PV = nRt$, where n is the number of moles present and R is the universal gas constant.

We may even extend this further when we remember that every mole of a substance contains Avogadro's number of molecules for a molecular gas or atoms for an atomic gas. We can now use an even more sophisticated formula for the Gas Laws as follows:

$$PV = nNkT$$

where N is Avogadro's number (6.025×10^{23}) and k a constant called Boltzman's constant, which is the same for every gas as long as it is in the ideal gas state.

It is interesting to note that the above relation is independent of the type of gas or mixture of gases. For instance, 1.5 moles of O_2, 8 moles of N_2, and .5 moles of CO_2 would have exactly the same volume at the same pressure and temperature as any one of these gases would if 10 moles of it alone were present.

We can now have our students calculate the Boltzman constant for an ideal gas by reminding them that at 273°K and standard pressure of 1.02×10^5 newtons/m², one mole (6.025×10^{23} molecules) would occupy 22.4 liters or $2.24 \times 10^{-2}\text{m}^3$.

$$PV = nNkT$$

$$k = \frac{P \times V}{n \times N \times T}$$

$$k = \frac{1.02 \times 10^5 \text{ newtons} \times 2.24 \times 10^{-2}\text{m}^3}{\frac{6.025 \times 10^{23}}{\text{mole}} \times 1 \text{ mole} \times 273°\text{K}} = 1.38 \times 10^{-23} \frac{\text{joules}}{°\text{K}}$$

The constant R in the form of the Gas Law usually used by physical chemists is really then the Boltzman constant times Avogadro's number.

$$1.38 \times 10^{-23} \frac{\text{joule}}{°\text{K}} \times 6.03 \times 10^{23} = 8.37 \frac{\text{joule}}{°\text{K-mole}}$$

WHAT'S IN A NAME

We have thus included in our model all the formally named laws without pointing them out. It was important to first emphasize the concept and build our working model. We might now, as a matter of interest and historical importance, relate the development of this model as it grew during the past few centuries and then give credit to the Daltons, Boyles, and Charleses by indicating the parts of the general gas formula bearing their names.

We can show that since equals divided by equals remain equal, we can always derive Charles's Law and Boyle's Law from this general gas relationship.

For instance, if $P_1V_1 = n_1NkT_1$ and $P_2V_2 = n_2NkT_2$, and we arrange our experiment so that the pressure and the number of moles remain constant (the gas might be contained in a cylinder with a close fitting piston), we are given a known initial volume and temperature and ask what the volume would be at some other temperature. The problem can be set up as follows:

$$\frac{P_1V_1 = n_1NkT_1}{P_2V_2 = n_2NkT_2}, \text{ since } P_1 = P_2, n_1 = n_2$$

We then end up with $V_1/V_1 = T_2/T_2$ and Charles's Law comes into being.

Part of where this argument is leading concerns what I believe is misleading in much teaching, due to a misplaced emphasis in many physics texts I have examined. As pointed out above, it is important that the relationships between the various parameters making up the ideal gas equation be understood, not what is Dalton's Law of Partial Pressures or what is Boyle's Law. We sometimes make the mistake of giving a name to something and think that we, therefore, understand it.

For instance, we are wont to explain the existence of liquids or solids with names like "cohesive forces" and the wetting of a solid by a liquid as due to an "adhesive force," but it must be understood we are only giving a name to something, not explaining it. Later we might talk about electrical forces and gravitational forces and, maybe, nuclear forces. Again we give names to poorly understood phenomena, but at least we have only those three forces, not the myriads we seem to have when we call these electrical effects between molecules adhesive, cohesive, Vander Waal's etc. At least the fundamental nature of the phenomenon ought to be finally tied down. Later we can ask the student to explain electrical attractions between apparently neutral particles.

WEIGHT AND MASS

As long as we are on the subject of names, it would be appropriate to discuss a case (as an example of many others) in which what we call something is very important. This is nowhere more exemplified than in the distinction between weight and mass. Laymen tend to use these terms almost synonymously and scientists perpetuate this error in their own handling of these terms.

The chemist talks about the atomic weights of elements when he really means the atomic mass. The chemist knows he is referring to the mass and not the weight of the elements, but the student is confused and will tend to remain confused unless the distinction is drawn early and sharply. We also help perpetuate this confusion when we talk about weight in mass units. We suggest that something weighs 3 kilograms. We really mean it weighs as much as 3 kg. weighs in the gravitational field near the surface of the earth, but all this is understood. It is not obvious to the student.

Mass is defined in several ways. In our first contact with mass, as in chemistry, we define the mass of a body as the quantity of matter it contains—i.e., the number of atomic mass units. Later, as we become familiar with the laws of dynamics, we can measure mass and define it as a quantity of inertia or an amount of resistance to change of motion, or we might measure it by the amount of attractive force it exerts on a given mass at some given distance. (We will distinguish between inertial and gravitational mass in a subsequent chapter.)

Weight, on the other hand, is the amount of force exerted on a given mass in the gravitational field of some other mass.

Mass is a property of matter which has nothing to do with the location of the object under consideration.

The weight of the object depends on where it is. An illustrative example might be suggested as follows:

Suppose you had one Kg of butter. If you weighed it on an equal arm balance with a standard one Kg mass, it would weigh as much as the standard Kg mass does near the earth's surface. If you weighed it with a spring scale calibrated in kilograms, it would still appear to weigh as much as the standard Kg does near the earth's surface.

Now if we took the Kg of butter, the standard Kg, and our scales to the moon to repeat our measurements, we would find some differences in our results. The kilogram of butter would still contain the same amount of matter in it. We can still spread the same number of bread slices with it. When we balanced it with the standard Kg on our equal arm balance, we would find that it weighed as much as the standard mass near the surface of the moon, but when we weigh both the butter and the standard Kg mass with our spring scale, we would find they both weigh only about 1/7 of what a Kg would weigh near surface of the earth. It is then necessary, in order to express weight, to use force units, and we will define force units later. In the M.K.S. system, we will use newtons.

The importance of using the required terminology at first, and I suggest always, cannot be overstressed. Later, when your students become knowledgeable and at ease, they may talk about atomic weight and something weighing so many kilograms as chemists do, but they will know they really mean something else; as long as they talk to others as sophisticated as they have become, there will then be no confusion.

It is true that sometimes we must present a simplified version of an explanation because the student's background is not yet ready for a more rigorous approach. It is necessary to point out, then, that it is not the complete story but will suffice only until more groundwork is laid. We must be careful, however, and never present a model that is wrong in the interest of oversimplification.

This is especially true in our system of education where, in most schools, biology, the most complicated of the sciences, is taught first and chemistry, which has considerably fewer unknown variables, is usually next. Physics, the most basic science and the one which can be taught most quantitatively because it is perhaps best understood, is then taught last. It would seem that a counter-current would be desirable, but then we would run into the other problem of having to teach that science which can be handled quantitatively and with most complete understanding when our student would be least able to handle it. If we teach science in packets labelled Biology, Chemistry, and Physics, it is perhaps best that we save physics for the last, when our students have attained a maturity permitting a better grasp of abstract principles, and have had time to acquire the skill in mathematics necessary for quantitative work.

Another alternative might be similar to the system in many European countries, where several sciences are taught concurrently at different levels as students progress through school. A third suggested approach might be to do away with labeling the science package and to design a four-year science course in which we intertwine the sciences we usually teach. We could hold some of the more sophisticated areas of chemistry until the student understands the underlying physical concepts and then apply this understanding to living systems.

Schools might also arrange their programs so that it is possible for selected students to elect biology in their freshman year, chemistry in their second year of high school, physics in their junior year, leaving the senior year free for "senior science." This would not be an advanced placement course but would be designed to give advanced students an opportunity to reexamine problems in biology with the background acquired in chemistry and physics. This last might be the most easily adopted procedure since it entails the least drastic revision in our current procedures.

4

Review of Mathematics

As long as algebra and geometry proceeded along separate paths, their advance was slow and their applications limited.

But when these sciences joined company, they drew from each other fresh vitality and thenceforward march on a rapid pace toward perfection.

JOSEPH LOUIS LAGRANGE

PERHAPS ONE OF THE MOST DIFFICULT PROBLEMS physics teachers face is the lack of mathematical proficiency on the part of their students. Even some students who are reported to be doing well in their math classes seem to have difficulties applying the principles they are supposed to have learned.

Students electing physics should have had courses in basic algebra and geometry. Some trigonometry would help, but you may be required to define a sine, cosine, or tangent for them and show them how to use these. If they have not previously been taught to use slide rules, it is worth your time to teach this skill and to encourage them to continue using their slide rules throughout the course. Certainly, it is helpful that your students come to you with some well-founded preparation in mathematics, but it is also true that a good quantitative physics course will make their mathematics come alive for them. What they had previously been taught as abstract operations now becomes a useful tool for more clearly understanding the physical world in which they live.

We have discussed, in the chapter on measurement, some aspects of quantitative techniques we need to know, i.e., significant figures, indirect measurements, etc. In accumulating and interpreting experimental data, we look for relationships. Sometimes this is more easily done by expressing our data in graphical form and analyzing the pictorial representation seen in the graph. For instance, much can be learned about kinematic and dynamic relationships by observing the slope of a displacement-time graph or the area under a velocity-time graph.

These will be discussed here and elaborated on further in subsequent chapters. It would be well to illustrate at this point, however, the derivation of a useful relationship both by experimental techniques and by inductive reasoning.

THE INVERSE SQUARE LAW

It is useful to establish the inverse square law early in the course because of its general application to many physical phenomena.

The experiment might involve a small light source and a photometer or a source of gamma radiation and a geiger counter. A set of data is accumulated and arranged in an orderly table as follows:

Distance from light source	*Intensity illumination in ft. candles*
1 foot	26
2 feet	10.5
3 feet	7.3
4 feet	6.3
5 feet	5.8

The data can then be represented in a graph. (See Figure 4–1.)

It is evident that intensity of illumination at some distance from the source is not proportional to the distance. Since the slope is evidently inverse, try to graph the illumination vs. the inverse of the distance. The constantly increasing slope of the resulting graph looks like an exponential relationship and another trial is suggested. (See Figure 4–2.) This time the illumination is plotted against the square of the inverse of the distance. This suggests a direct relationship (I vs. $1/d^2$), and the fact that the line crosses the I line above the origin indicates a constant background. (See Figure 4–3.)

The inverse square law can be arrived at by arguing from a geometrical representation as follows:

Visualize a point source of radiation which is constantly emitting particles. There is no preferred direction and, therefore, if the radiation intensity is great (or the number of particles traveling through a cross-section of space is large), the chances are that the number of particles emitted in any direction would be the same. (See Figure 4–4.)

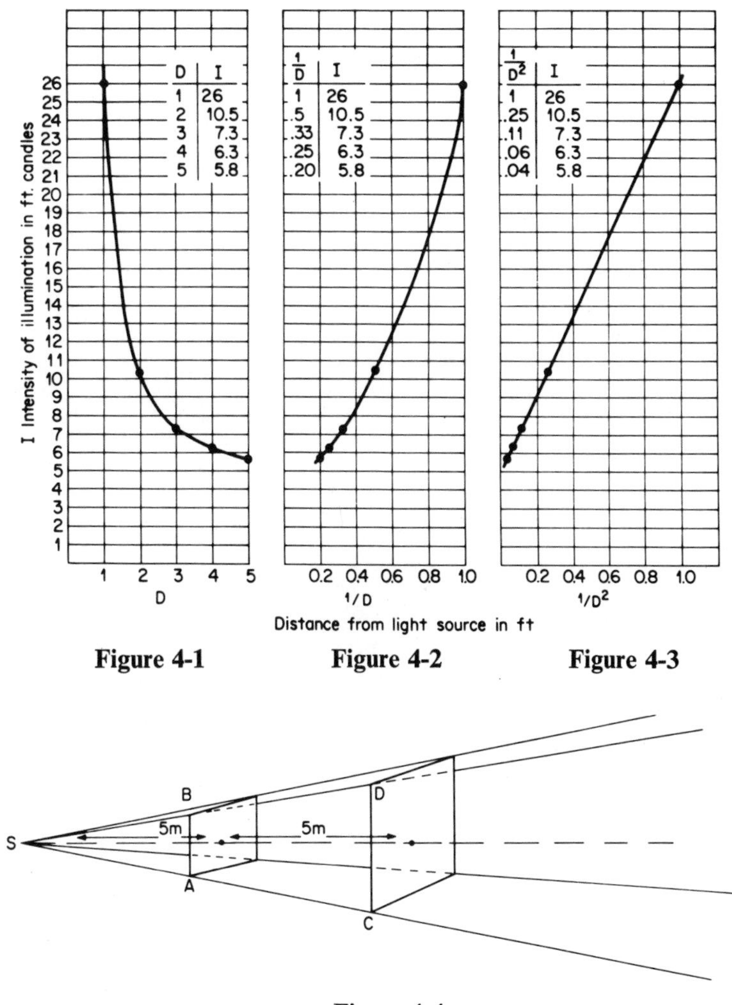

Figure 4-1 Figure 4-2 Figure 4-3

Figure 4-4

Note that the second screen is placed twice the distance from the point source of radiation as the first screen. Note also that all the particles leaving point P traveling inside the solid angle will go through the two screens. As is evident from the figure, ABS is similar to CDS. This is true of all the triangles making up the sides of the figure. Since the larger base is twice as far from the apex S as is the smaller base, it must be twice as long and twice as wide, and the area of the larger screen is four times that of the smaller and closer screen. Since intensity is an expression of the radiation received per unit area and the total radiation received by both screens is the same, it follows that the intensity of radiation received at a given distance from a source is inversely proportional to the distance from that source or $I = k/D^2$.

Another geometrical argument for the inverse square law might be presented as follows (see Figure 4–5.):

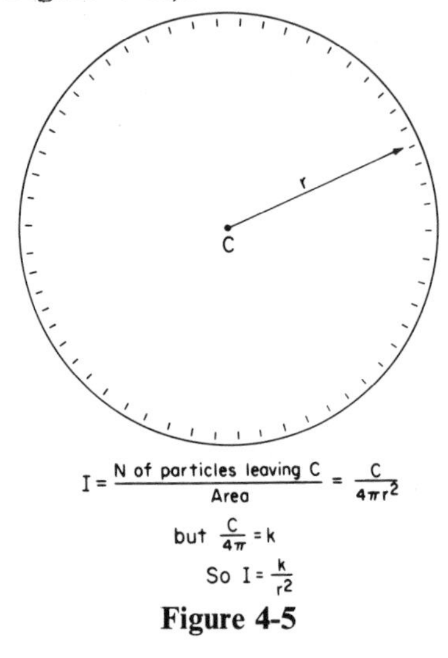

Figure 4-5

Visualize the same point source radiating evenly in all directions. Imagine a spherical surface whose center is at the point of radiation and whose radius is R. Since the area of a sphere is $4\pi R^2$ and all the radiation must go through their surfaces, then the intensity of radiation received at each surface is the same amount of radiation per area. We can again find that the intensity of radiation received at a given distance from a source is inversely proportional to the distance for that source. In this case $I = C/4\pi R^2$. Since $C/4\pi$ is a constant, then $I = k/R^2$.

We can use these illustrations to demonstrate an analytical approach involving a geometrical visualization. This leads to an algebraic relationship which then becomes a tool for predicting and extrapolating.

An example of such a useful extrapolation might be indicated in the calculation of the distance of stars too far to measure by parallax.

Exercises in measurements by triangulation and parallax were discussed in Chapter 2. Astronomers tell us that they can measure the distance of stars to within 500 light years by parallax. We are also informed that stars of different types emit radiation at different energy levels. Since stars have been placed in only seven different categories (for type of light that they emit), we can use this classification for measuring their distances from earth.

Stars are categorized as O, B, A, F, G, K, and M stars. Perhaps you might amuse your students by reciting the following mnemonic "Oh, Be A Fine Girl; Kiss Me."

O-stars are massive and hot and emit light mainly from blue to ultra-violet. M-stars are small and cool, emitting most of their radiation at the red end of the spectrum. Our sun is a G-type of star which is somewhere between the other two in mass and emits most of its light in the yellow region of the spectrum.

In the inverse square relationship previously derived, $I = k/d^2$, the constant (k) would refer to the amount of radiation emitted by the source. Observations indicate that all O-type stars emit the same quantity of radiation and that any B-type star gives off exactly the same amount of light per unit time as any other B-type star. When using the inverse square law to determine the amount of light received at a given distance from such a star, we can in our calculations handle the two similar stars at two different distances exactly the same as we would a single source moved to two different positions. Thus, if $k = I\,d^2$, then $I_1 d_1^2 = I_2 d_2^2 = I_3 d_3^2$, etc. It then becomes possible to measure the distance of a "fixed" star too far away to observe any parallactic shifting during the course of the year by comparing the light received from it (which can be measured) and the light received from a similar type of nearby star whose distance has been established by parallax and triangulation. Thus, since $I_n d_n^2 = I_f d_f^2$, then $d_f = \sqrt{I_n d_n^2 / I_f}$.

SCALING

Though scaling is not necessary for exploring the main stream of physics, it offers many useful ideas and furnishes a good groundwork for quantitative reasoning. We see pseudo-science fiction movies about mutant spiders grown to over one hundred times their natural size. Aside from the biological implications of such sudden changes, could such monsters physically exist? We know that the strength of its members must change with its cross-sectional area, while its mass (assuming a constant density for spider flesh) must increase with its volume. These monsters will be ten thousand times as strong as their normal ancestors but must carry about a million times their normal weight. They won't even be able to lift their bellies off the ground.

A story to help illustrate this and other concepts might be related as follows:

Our astronauts landed on Venus and observed a welcoming committee approaching in the distance. They looked like well-proportioned earth-women. As they approached, our astronauts found that they had misjudged their size (there were no familiar landscape objects with which to compare). The women turned out to be about fifty feet tall. The leader of our earth group turned to his navigator and chided him for putting them on the wrong planet. How did he know he was not on Venus?

The women were ten times the linear dimensions of earth women. They were, therefore, one hundred times as strong with one thousand times the mass. If they were on an earth size planet, they would be carrying one thousand times the weight of earth beings with only one hundred times the strength. They must have landed on a planet with approximately one tenth the gravitational field of the earth. Maybe they were on the far side of the moon.

Other biological implications might be cited. There are limitations of smallness in warm-blooded animals, as in the shrew and humming birds, whose surface to volume ratio is such as to require almost constant food ingestion in order to replace the heat lost through surface radiation. Limitations in largeness are illustrated by the whale who could not possibly adapt to live on land and maintain its present gigantic size, or the final extinction of the dinasours who became too large to compete successfully with their environment.

Another interesting effect of the scaling factor is its effect on the races of man. Note the relatively spherical shape of the Eskimo people whose surface area must be small, relative to body volume, to protect them from losing too much body heat, while the people living in hot climates, like equatorial Africa, have generally evolved into slim, long-limbed types, thus permitting a large surface area for radiating away excess body heat.

A knowledge and understanding of scaling factors is, of course, important to engineers and model builders. A scaled down version of a model airplane tested in a wind tunnel must be scaled according to the known factors in order to extrapolate the behavior of a full-size airplane.

GRAPHICAL ANALYSIS

Before introducing your students to the study of kinematics and dynamics, it is best to cite examples from their own experiences.

When we ride in a car, we observe the speed with which we are traveling and express this speed in miles per hour or units of distance divided by units of time. If we want to know how far we have gone in a certain interval of time, we merely multiply the speed by the interval of time to arrive at the distance.

$$\frac{\text{miles}}{\text{hr.}} \times \text{hrs.} = \text{miles}$$

Please note that units can be handled algebraically just like numbers and letters. Another way to handle this problem is to do it graphically. Let us assume that we have been traveling at a rate of 50 miles for 8

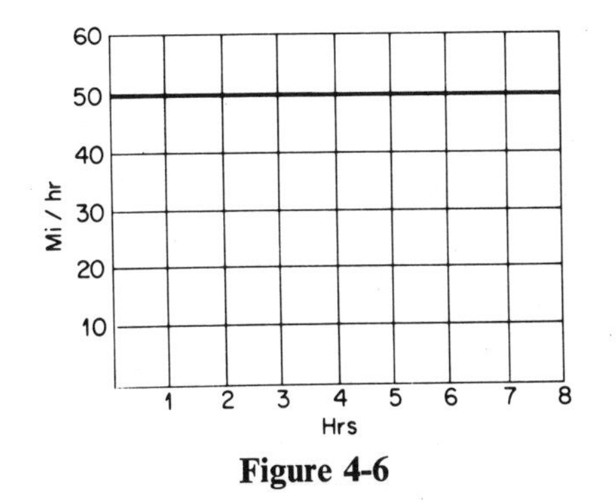

Figure 4-6

hours. A graph plotting the rate against time would look like Figure 4–6.

Suppose you wish to calculate how far you traveled. 8 hr. × 50 mi./hr. equals 400 miles, but how can we tell from the graph? The area under our line must represent the distance since it is a rectangle with dimensions of 50 mi./hr. and 8 hrs., whose area then is 400 miles, the product of its length and height.

Now, suppose we started measuring the distance from the time we started on our journey. We start, of course, from rest. It would be simpler for our problem to jump into a car already moving at 50 miles per hour but a little hard on us physically.

Let us say that in the first hour we gradually and smoothly attained a speed of 50 miles and after that we traveled at the same speed for seven hours. How far did we travel? Let us graph this trip as before. (See Figure 4-7.)

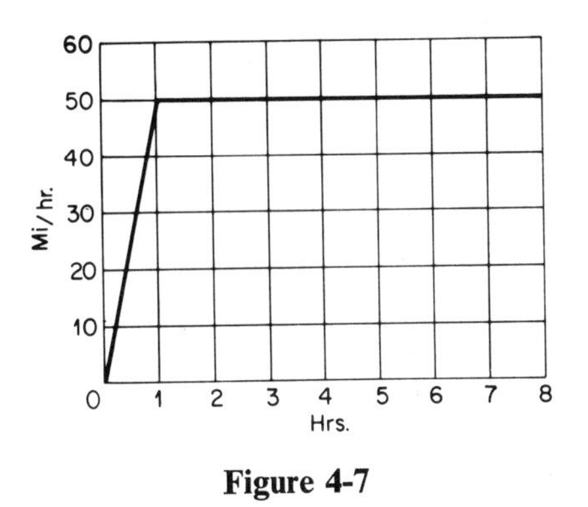

Figure 4-7

The area under the graph represents the distance traveled and consists of a triangle for the first hour and a rectangle for the next seven hours.

The area of the triangle is one-half the base times the altitude or $\frac{1}{2} \times$ 1 hr. × 50 mi/hr = 25 miles. The area for the next seven hours is 50 miles/hr

× 7 hrs. or 350 miles. The total distance traveled would then be 375 miles.

Another way to do this problem, of course, is to say that if we changed our speed gradually and evenly from 0 → 50 miles for the first hour, then our average speed for the first hour was 25 miles per hour, and continue the problem by multiplying average speeds for different intervals of time and multiplying by that interval of time to get the distance traveled. We would, of course, arrive at the same result, and it might be easier in such a simple problem than constructing and analyzing a graph. Suppose, however, that during the trip we were forced to speed up and slow down, start and stop. We carefully recorded our speed at regular and frequent intervals of time, and after eight hours our graph might look like the one depicted in Figure 4-8.

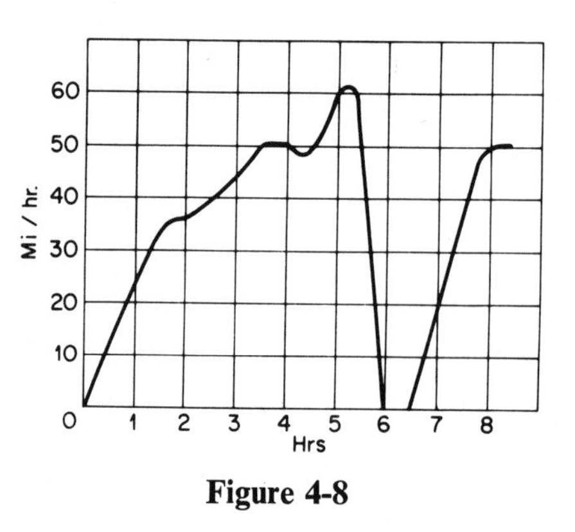

Figure 4-8

We can see that our speed did not increase smoothly and evenly for the first hour, that we did not maintain an even speed for the rest of our trip, and that we even stopped for about thirty minutes at one part of the trip. How far did we travel?

Using the area under the graph to indicate the distance would be a comparatively easy way to arrive at a solution. Several methods of determining the area under an irregular line suggest themselves to us.

1. We might divide the area into regular parallelograms and triangles to solve individually and sum up.
2. We can rule off small squares of known area, count the number of these we can fit into the large irregular area, subdividing them into smaller and smaller areas near the edges where they do not fit, until we arrive at an acceptable estimate of the distance.
3. Another neat method is a less direct approach. If we copied our graph on a piece of paper of an even texture, we could cut out the area we are measuring and also cut out an even rectangle of known

area. We could then arrive at the unknown area by weighing these two pieces of paper since the areas would be proportional to the weight.

We can also use this velocity-time graph to determine our acceleration. During the eight hours recorded, we started from zero miles per hour and were moving at 50 miles per hour after eight hours of travel. Our average acceleration, therefore, was 50 mi/hr/8 hrs = 6.2 mi/hr^2. During this interval, however, we had speeded up, moved at constant speed, slowed down, etc.—the acceleration was certainly not constant. For instance, what was the acceleration between the fifth and sixth hour?

Perhaps we ought to stop here for a moment and introduce a convenient notation. The delta notation (ΔV) means *change in velocity;* that is, the difference between the final velocity and the initial velocity. Acceleration might, therefore, be defined as $\Delta V/\Delta T$ or $(V_f - V_1)/(T_f - T_1)$. In the problem suggested above, i.e., the acceleration during the fifth and sixth hour, we would set it up as follows:

$$A = \frac{\Delta V}{\Delta T} = \frac{V_f - V_1}{T_f - T_1} = \frac{0 \text{ mi/hr} - 60 \text{ mi/hr}}{6 \text{ hr} - 5 \text{ hr}} = -60 \text{ mi/hr}^2$$

The average acceleration is different during this one hour period than it is for some other one hour period. It is not the same as the average acceleration during the entire eight hour period previously determined. How then can the acceleration be determined at any instant of time? Well, we will keep zeroing in until we determine $\Delta V/\Delta T$ as ΔT approaches 0. For instance, in the graph in Figure 4-9, one would determine the shape of the line showing the acceleration during the first and second minute of travel to be:

$$\frac{\Delta V}{\Delta T} = \frac{V_f - V_1}{T_f - T_1} = \frac{47 \text{ mi/hr} - 11 \text{ mi/hr}}{2 \text{ min} - 1 \text{ min}} = \frac{36 \text{ mi/hr}}{\text{min}}$$

Similarly, the average acceleration between 1.5 min. and 2.5 min. after the start of travel would be:

$$\frac{36 \text{ mi/hr} - 26 \text{ mi/hr}}{1 \text{ min}} = \frac{10 \text{ mi/hr}}{\text{min}}$$

To obtain the instantaneous acceleration at exactly two minutes after the start of the journey, one measures the slope of the tangent at that point and finds it to be about 4 mi/hr/min.

Using areas and slopes of graphs to determine relationships will be explored further in subsequent chapters. Familiarizing your students with these techniques as early as possible will help make problem solving much easier. The easier it is for students to solve problems, the easier it will be for you to maintain their interest in physics.

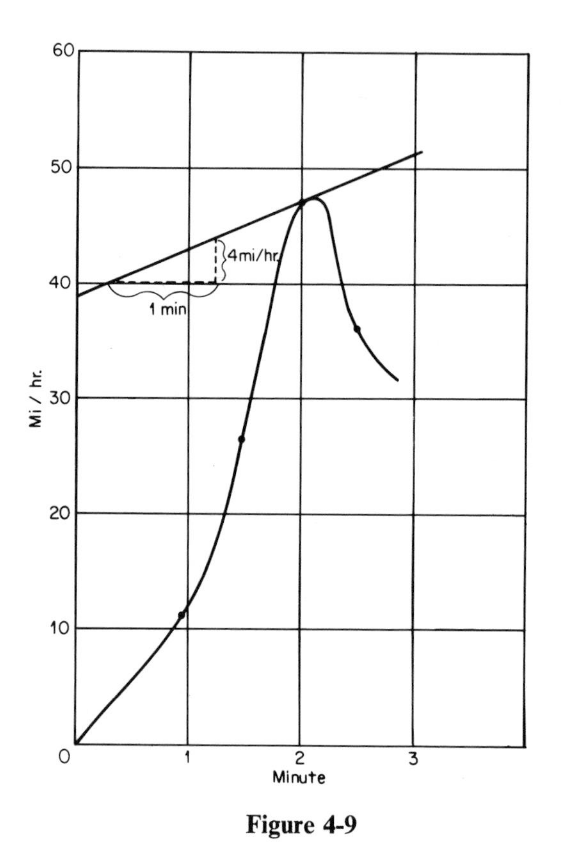

Figure 4-9

DIMENSIONAL ANALYSIS

Students should be taught from the beginning to include units in the data used for solving problems and to manipulate these units algebraically to make sure that they end up with the units required in the solution.

For instance, let us take a very simple case (perhaps too obvious, but it will serve to illustrate). Suppose we wish to know how long it will take to drive 50 miles if we can maintain an average rate of 40 mi/hr. How do we set up the problem for solution? We are given distance and speed and we want the time. We can see by inspection that distance divided by speed equals time. $D/V = D/(D/T) = D \times (T/D) = T$, therefore 50 mi/40 mi/hr $= 1\frac{1}{4}$ hrs.

Later we will find that if we send a charged particle through an electric field which applies a downward force on the particle, while at the same time it is subjected to a magnetic field which applies an equal but upward force, the particle will go through undeviated. We can ascertain its velocity by dividing the electric field by the magnetic field. The fact that $v = E/B$ is not intuitively obvious but can be justified by dimensional analysis as follows:

The electric field (E) is defined as force per unit charge—in the M.K.S. system, newtons/coulomb. The magnetic field is measured in terms of newtons per ampere-meter.

$$E/B = \frac{\text{newtons/coulomb}}{\text{newton/amp-meter}} = \frac{\text{ampere-meter}}{\text{coulomb}}$$

$$= \frac{\frac{\text{coulombs}}{\text{sec}} \times \text{meter}}{\text{coulomb}} = \frac{\text{meters}}{\text{sec}}, \text{ therefore}$$

E/V has the units of velocity.

Another example before we finish this section might be appropriate. In Chapter 3, we arrived at a simple expression for the ideal gas law, $PV = RT$. Since RT, the temperature times the ideal gas constant, is a measure of the energy contained in a given amount of gas, then $P \times V$ ought to have the units of energy. This can be easily demonstrated since (P) pressure equals Force per unit area and $F/A \times V = F/A \times A \times D = F \times D$, which is a measure of work or energy.

Finally, the following is an illustration of the use of graphical analysis for the derivation of a well-known kinematic relationship from experimental evidence.

A MATHEMATICAL ANALYSIS OF AN EXPERIMENT

Early in the year, we attempt to teach kinematics to our physics classes and arrive at the relationship $D = \frac{1}{2} AT^2$. While this can be done both algebraically and graphically, I believe that kinematics could become more meaningful if this formula were to be experimentally derived and then supplemented by the other treatments. It also provides a means of introducing, early in the course, the experimental method and the analysis of experimental results.

NOTE: The laboratory exercise described here is for two periods, and I have found that it may require another class period for review. The time spent is, I believe, well worthwhile in the better understanding of the kinematic relationship.

For my experimental approach, I use the bell clapper timers and tapes designed for the PSSC laboratories on kinematics and dynamics (these may be purchased from many sources. As a matter of fact, this laboratory exercise grew out of an extension of PSSC lab 1-5.

Here is how the equipment works: A weight is attached to the paper tape and is allowed to fall to the floor while the timer (Figure 4-10) makes its carbon marks on the tape. (A small C clamp is a convenient weight to attach to the paper tape.) The tape will then show the characteristic acceleration pattern of dots (Figure 4-11A). The student is then asked to plot Displacement v. Time. The steps are as follows: 1. For his time interval, the student can assume that the period of the bell clapper was constant and that one, two, or three claps might constitute one interval of time. He can then lay his tape down along the Y-axis of the graph (Figure 4-11B), rule off the T-axis in the regular time intervals, and mark off the points in his graph as shown in line A of Figure 4-11B.

Alternative timer: If you don't have a bell clapper timer, you can make one. See instructions at the end of this section.

2. It is evident from the curve that D is not linearly proportional to T. The curve, however, looks as if it might be parabolic, and this "hunch" can be explored by plotting D v. T^2. This is done on the same graph by leaving the first interval where it is. The dot for the second interval is moved to 4 (2 squared), the dot for the third interval is moved to 9 (3 squared), the dot for the fourth interval is moved to 16, and so on. A line (line B on Figure 4-11B) connecting these dots appears to be a straight line and, therefore, D is linearly proportional to T^2. The slope of line K is D/T^2 and thus $D = KT^2$.

3. The experiment can be further extended by suggesting that K may bear some relationship to acceleration. The acceleration is discovered by plotting the Velocity against Time. The same T-axis is used as before, and the tape is laid against the Y-axis as before. This time, however, it is pulled down each time we plot succeeding displacements to place each succeeding dot on $Y = 0$. By doing this, we plot changes in distance per unit of Time, rather than distance in total time. The resulting straight line (line C on Figure 4-11B) indicates that V is linearly proportional to T. The slope V/T is the acceleration, and thus $V = AT$. The numerical value of this slope (A) which is in units of V/T or $D/T/T$, can be compared with the numerical value of the slope (K) in the previous line (K is in dimensions of D/T^2). It will be then discovered that within experimental error, $K = \frac{1}{2} A$ and thus $\frac{1}{2} A$ can be substituted for K in the formula $D = KT^2$, to give $D = \frac{1}{2} AT^2$.

Having thus first derived the formula by experiment, we can then bring in the following algebraic and graphic derivations to supplement the experiment. This gives confidence in the use of mathematics.

Algebraic

The derivation can be arrived at by algebraic manipulation of the definitions as follows:

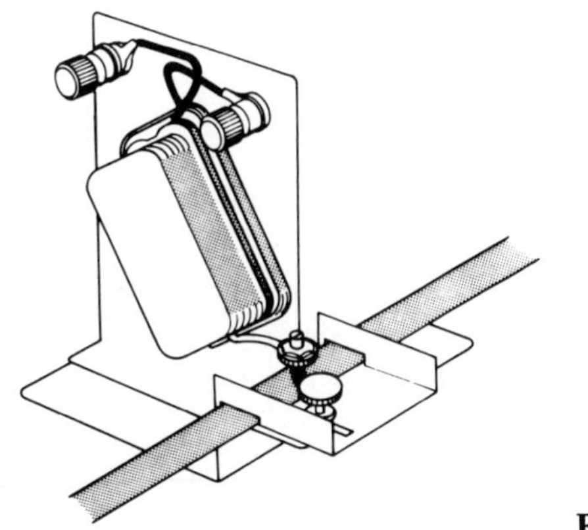

Figure 4-10

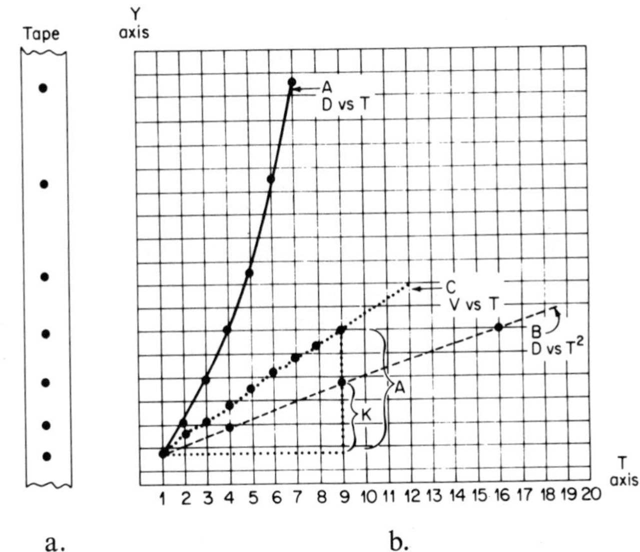

Figure 4-11

$\bar{V} = \frac{D}{T}$ (Average velocity is total change in displacement per interval of time.)

$A = \frac{V_f}{T}$ (Acceleration is the total change in velocity per unit of time and the final velocity is the total change in velocity if the object starts from rest.)

but

$\bar{V} = \frac{1}{2} V_f$ (The average velocity is one-half the final velocity if the object starts from rest and accelerates at a constant rate to the final velocity.)

Therefore:

$$D = \overline{V}T$$
$$D = \tfrac{1}{2}\, V_f T \text{ and } V_f = AT$$
$$D = \tfrac{1}{2}\, AT \cdot T = \tfrac{1}{2}\, AT^2$$

Graphic

Another useful presentation is to derive the relationship graphically by graphing V v. T as the V increases regularly with T, as is shown in Figure 4-12A. (Note comparison with line C, Figure 4-11B.)

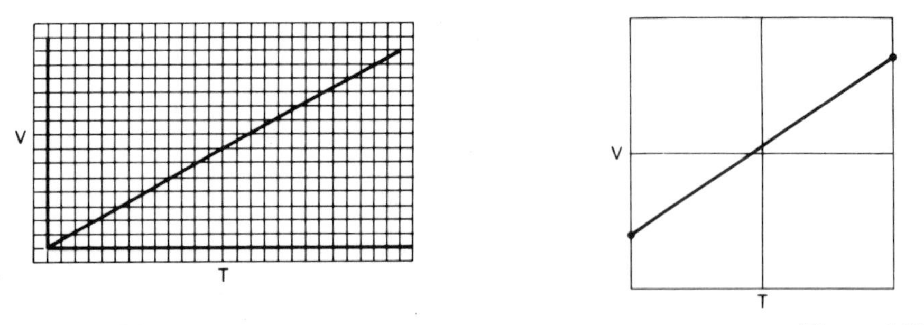

Figure 4-12A **Figure 4-12B**

Since the area under a velocity/time graph represents distance, the triangle seen in the graph in Figure 4-12A signifies $D = \tfrac{1}{2}\, V_f T$. The slope of the velocity/time with constant acceleration will be constant and the slope $A = V_f/T$ and $V_f = AT$. Substituting for V_f in $D = \tfrac{1}{2}\, V_f T$, $D = \tfrac{1}{2}\, AT^2$.

The presentation can be extended by suggesting V_0 is not equal to zero. The graph of V v. T then looks like Figure 4-12B.

The graph shown indicates an area made up of a rectangle and a triangle so that $D = V_0 t + \tfrac{1}{2}\, AT^2$. One more extension would be to ask the student to imagine starting his observations at $T = 0$ when the object whose action he is observing is already at some distance D_0 from him, then $D = D_0 + V_0 t + \tfrac{1}{2}\, AT^2$.

Alternative timer: Here's how you can make a timer from a doorbell from which the gong has been removed, a microscope slide, some thumb tacks, and wood. Construct the basic timer as shown in Figure 4-13 (it can be operated by a pair of flashlight batteries). Then operate as follows:

The recording tape (strip of paper about $\tfrac{1}{4}$ inch wide) is drawn under the vibrating bell clapper over a hard surface (the glass microscope slide) while a piece of carbon paper is moved around between the bell clapper and the moving tape, in order to present a fresh surface for marking on the paper.

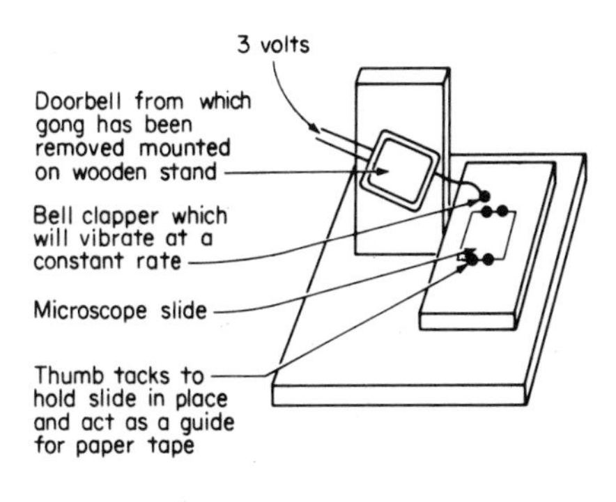

Figure 4-13

DERIVATION OF A RELATIVITY FORMULA

A problem illustrating the use of algebraic skills to arrive at a general relationship is proposed as follows:

A boat captain makes regular trips down river to a certain town and then returns. His boat always travels at the same speed relative to the water. He has a choice of keeping his boat in a canal, connecting the two towns, in which no current is flowing, or keeping his boat in a river running parallel to the canal. The river has a constant current flowing. Will there be a time difference in going back and forth in the still waters of the canal or going downstream with the current in the river and back against the current?

We might set up our problem as follows: let D (the total distance of the trip) equal $2\,d$ (the distance between the towns).

Let $\vec{U}$ = stream velocity

$\vec{V}$ = vessel velocity with respect to water

T_1 = time of trip with $\vec{U} = 0$

T_2 = time of trip with $\vec{U} > 0$

Then $\vec{V} = \dfrac{\vec{D}}{T_1}$ and $T_1 = \dfrac{\vec{D}}{\vec{V}}$

$$\text{But } T_2 = \frac{d}{\vec{V} + U} + \frac{d}{V - \vec{U}}$$

(The total time of the trip is the time to go the downstream distance when the speed of the current is added to the speed of the boat plus the time to go upstream when the boat is opposing the current.)

Simplify by finding the least common denominator:

$$T_2 = \frac{d\,(V - U) + d\,(V + U)}{V^2 - U^2}$$

$$T_2 = \frac{2\,dV}{V^2 - U^2} = \frac{DV/V^2}{V^2 - U^2/V^2}$$

$$T_2 = \frac{D/V}{1 - U^2/V^2} = \frac{T_1}{1 - U^2/V^2}$$

Now we can see from the formula we derived that if U is very small compared to V, $1 - U^2/V^2$ is still $\cong 1$ and $T_1 \cong T_2$, but as U approaches V in magnitude, T_2 (the duration of time to make the trip when a current is flowing) becomes very large compared to T_1.

The technique of setting up a general equation to use for predicting the effect of a variable is a very useful analytical tool in science.

SAMPLE PROBLEMS

A group of bank robbers were fleeing from the police at a rate of 70 miles per hour. The police were able to maintain an average speed of 80 mi./hr. At exactly noon on the day in question, the robbers were 100 miles from a state border while the police, who had just been alerted to their escape route, were 5 miles behind them.

Can the robbers be intercepted before they reach the border?

At what time would this interception take place?

Where would the interception take place?

There are several methods of solving this problem. A graphical solution would serve as an illustration of one method of doing it. We can draw a distance-time graph in which $T = 0$ (12 noon), the police are at the origin, the robbers are 5 miles up the Y axis, and the border is 100 miles up the Y axis. The slopes of the graphs of the robber and police cars are determined by the given speeds (distance/time). Where these two lines intercept would designate the point of interception in distance and time as follows (see Figure 4-14):

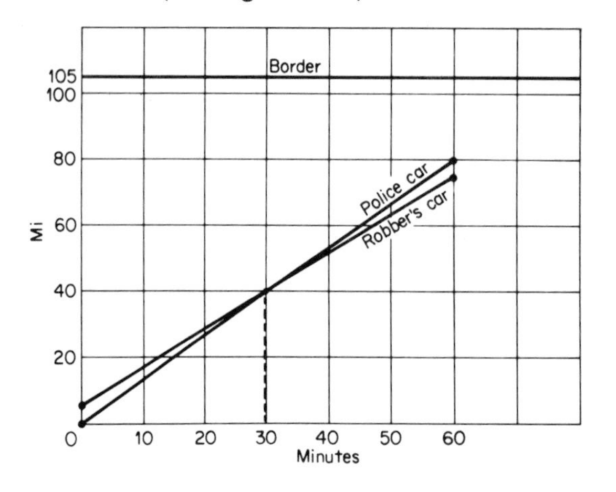

Figure 4-14

An inspection of the graph would indicate that the robbers would be caught 65 miles from the border at 12:30 P.M.

Another way to do this same problem is to recall that the police would intercept the robbers when they are both the same distance from the same point at the same time. We can take the point from which we are measuring as the place where the police car is at 12 o'clock.

We know that the

$$D = D_0 + \Delta D \text{ and } \Delta D = R\,\Delta T$$

That is: The distance moved during an interval ΔT is the rate (R) times the interval (ΔT), and the total distance is the original distance D_0 plus ΔD (the change in distance).

For the police car,

$$D = 80 \text{ mi/hr} \times T$$

For the robbers' car,

$$D = 5 \text{ mi} + 70 \text{ mi/hr} \times T$$

We equate these,

$$80 \frac{\text{mi}}{\text{hr}} \times T \text{ hrs} = 5 \text{ mi} + \frac{70 \text{ mi}}{\text{hr}} \times T \text{ hrs}$$

$$80\,T \text{ mi} = 5 \text{ mi} + 70\,T \text{ mi}$$

$$80\,T = 5 + 70\,T$$

$$10\,T = 5$$

$$T = \tfrac{1}{2} \text{ hr}$$

In 30 minutes, the police car had gone 40 miles and the interception would take place while they are still 65 miles from the state border.

A baseball is dropped from the roof of a building 40 meters above the ground. As it falls to the ground under the influence of the earth's gravitational attraction, it will accelerate at a rate of about 10 meters/sec^2.

How long will it take to hit the ground from the time it is dropped, and how fast will it be moving at the time it hits the ground?

We can choose to solve this problem graphically by constructing a velocity-time graph in which the velocity line shows a constant slope of $10 \frac{\text{meter/sec}}{\text{sec}}$ and it starts at the origin when $T = 0$. (See Figure 4-15.)

We know that the distance the baseball fell would be represented by the area under this velocity-time graph. We know that this area should be 40 meters. We could then inspect the graph and discover the 40 meters would be represented by the area of the triangle whose altitude is about 28 meters/sec and whose base is 2.8 + seconds. (Each square has an area representing 2.5 meters/sec × .2sec or .5 meters.) However, we had already

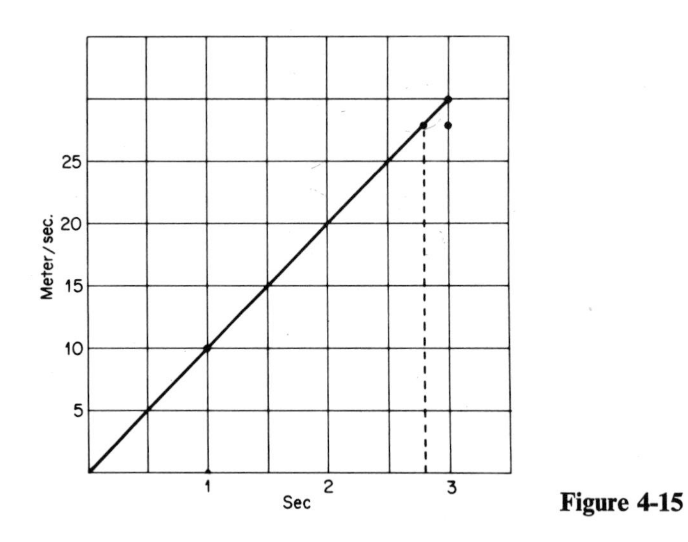

Figure 4-15

derived some useful kinematic formulas which we can use to solve this problem. We know that $D = \frac{1}{2} at^2$, therefore $40 = \frac{1}{2} \cdot 10 \cdot t^2$.

$$t^2 = 8$$

$$t = \sqrt{8} \text{ or approx. } 2.8 \text{ sec}$$

Since

$$V = \text{at}$$

$$V = 10 \times 2.8 \text{ or } 28 \text{ meter/sec}$$

or $V^2 = 2\,gh$

$$V^2 = 2 \cdot 10 \cdot 40$$

$$V^2 = 800$$

$$V = 28 \text{ meter/sec}$$

5

Vector Arithmetic

PROBABLY THE FIRST DIFFICULTY most students have in an introductory physics course is the concept of vectors. They have done scalar arithmetic all their lives and have learned to handle problems in scalar arithmetic with some degree of ease and confidence. Yet, for some students, a lack of confidence in handling problems involving vector quantities remains a constant deterrent to their complete understanding of physical relationships.

I recall an instance when I had occasion to have a physicist from a neighboring university visit us one Saturday morning near the beginning of the year to deliver a lecture to a group of bright high school physics students. His topic was "Nuclear Particles," and he did an excellent job of summarizing for his young audience what was then known about the thirty or so particles he discussed. My students were most attentive and after the lecture one of the young ladies in the group went over to him and said, "I enjoyed your lecture very much; I think physics is a fascinating subject. What I have trouble with is vectors."

Physics is sometimes described as "An Exact Science." The layman uses words like speed and velocity, weight and mass, etc., with no real appreciation for their meaning. Our students must learn to examine words for their exact meaning.

Ideas can be expressed clearly and precisely only if the meaning of the words used is clearly defined. Words should then be used only where they are appropriate or necessary. Everybody knows the meaning of the word *distance*. If you walk around the block, you cover a distance that can be measured by the number of steps you take times the average length of each step. However, if you end up where you started, your net *displacement* is zero. Displacement is a measure of distance moved in a given direction. If you walk all over town but finally end up 100 feet south of where you started, you may have walked a great distance but your total displacement is only "one hundred feet south."

Distance, which implies only the numerical magnitude of the trip in the above sense, is called a *scalar* quantity. Displacement, a quantity that specifies both magnitude and direction, is called a vector quantity.

Our students will come across many different kinds of vector and scalar quantities in their study of physics. They must learn to recognize the difference between vectors and scalars. Concentrating on the definitions of distance and displacement as I have in the short sample lesson above is most useful since once the idea is grasped, the recognition of other vectors becomes much easier. All the vector quantities we need to know in an introductory physics course are really the basic displacement vector operated on by one or more scalars. When a vector, which has both magnitude and direction, is multiplied or divided by a scalar, it can only have its magnitude affected while its direction must remain constant. Hence, the resultant of such an operation must remain a vector.

Some examples which illustrate this idea are the following:

Velocity must be a vector since it is defined as displacement divided by time and time is a scalar. Speed, on the other hand, must be a scalar since it is defined as distance per unit time and both distance and time are scalars.

Acceleration must, therefore, also be a vector if it is defined as displacement per time squared. I believe I can give this explanation a little more clarity by adopting the following scheme.

A number of scalar quantities will be listed and then various vector quantities will be shown to be simply displacement ($\vec{D}$) operated on by one or more of these basic scalars.

SCALARS

Time	(t)
Distance	(d)
Mass	(m)
Charge	(q)

VECTORS

Displacement	$\vec{d}$
Velocity	$\vec{v} = \vec{d}/t$
Acceleration	$\vec{a} = \vec{d}/t^2$

Force $\vec{f} = m\vec{a} = m\frac{\vec{d}}{t^2}$

Momentum $\vec{p} = m\vec{v} = m\vec{d}/t$

Electric field $\vec{E} = \vec{f}/q = \frac{m\vec{a}}{q} = \frac{m\vec{d}}{qt^2}$

Of course we have not included all the vector and scalar quantities, but I believe there are sufficient examples to illustrate the point I wish to make.

There are occasions when the product of a vector multiplied by another vector has physical significance. For instance, when we multiply a displacement by the force acting in the direction of the displacement, we obtain the magnitude of the work or the energy transferred by that force. Since only the component of the total force acting in the direction of the displacement contributes to the work done, then:

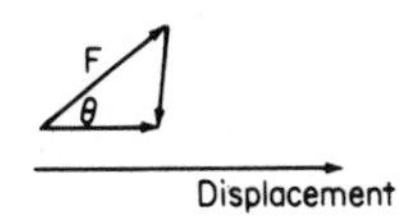

Work $= \vec{F} \cdot \vec{D} \cos \theta$

This is called a dot product and is a scalar quantity. One way to remember this is to think of a product of two linear dimensions determining an area. An area has magnitude but has no directional qualities parallel to its surface.

Another type of vector multiplication gives us a vector product or, as it is called, the cross product, which has both direction and magnitude. The direction is perpendicular to the surface of the plane determined by the two vectors being multiplied and a convention helps us determine whether this direction is up or down.

For instance, suppose we wish to determine the direction and magnitude of the torque caused by the tangential component of force applied at a given displacement from the center of rotation. The convention would have us set up the equation as follows:

$\vec{R} \times \vec{F} \sin \theta =$ torque

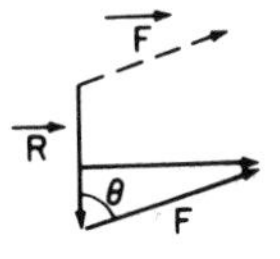

$\vec{F} \sin \theta$ is evidently the tangential component of the applied force and the direction is found by a right hand screw rule, that is, the direction a screw would move if you turned it from $\vec{R}$ to $\vec{F}$. Another way would be to open your right hand with your thumb held out at a right angle to your fingers and your palm facing in the direction from R to F. Your thumb then points in the direction of the torque. Obviously, the order of multiplying vectors to obtain the cross product is important since $\vec{F} \times \vec{R} \sin \theta$ would be in the opposite direction, your palm would point from $F \rightarrow R$ and your thumb would be reversed.

In adding and subtracting vectors, there are very simple rules to follow. Caution should be made here against teaching what is called the parallelogram method of adding vectors. It is limited to two vectors and may even be confusing to some students who may not really get to understand the properties of vectors if they are limited to this method.

Vectors are simply added by placing the tail of one on the head of the other, continuing this in any convenient order and finally determining the magnitude and direction of the resultant by drawing the vector whose tail is on the tail of the first vector and its head on the head of the last vector added.

Subtracting vectors is also as simple. The same algebraic rules that apply to scalars also apply to vector quantities. That is, if $\vec{A} - \vec{B} = \vec{R}$, then $\vec{A} = \vec{R} + \vec{B}$ or $\vec{A} + (-\vec{B}) = \vec{R}$.

A great deal of interest has been expressed in programmed learning. Whether or not this is all that some claim it to be is a moot question. It has been my experience that many students do not have sufficient self-discipline to work without continuing supervision, supplementary lectures, demonstrations by the teacher and, of course, supervised laboratory work. I have, however, experimented with short programs for enforcing specific concepts. These, I found, most students have the patience to work with and use for review, make-up work due to absence, or for some to move further on, if they wish, at their own pace when they become impatient with the progress of the class.

As an example of such short and specific single-concept programs, I append below a program I wrote to help my students understand how to work with vector quantities.

VECTOR PROGRAM

Cover all the questions following the one that is being read. After answering the question, lift the covering paper and check your result. If your answer is correct, go on to the next question. If your answer is wrong, recover the succeeding questions and reexamine the question you did incorrectly to ascertain what you did wrong and what should be done to obtain the correct answer.

1. Adding and subtracting scalar quantities consists simply of adding and subtracting the magnitudes of these quantities. For instance, 5 ft. + 2 ft. = 7 ft., 8 ft. − 3 ft. = _?, 4 seconds + 2 seconds = _?, 6 ft./sec + 2 ft./sec = _?

5 ft
6 seconds
8 ft/sec

2. Vectors differ from scalar quantities in that they indicate direction as well as magnitude. Instead of using numbers, arrows drawn to scale are used in which the arrow points in the direction indicated and the magnitude of the vector is depicted by the length of the arrow. If north is the direction toward the top of the page, and one inch equals 2 miles, draw the vector which represents a displacement of three miles northeast.

3. Vectors may be added by placing the tail of one on the head of the other and then drawing the resultant vector from the tail of the first to the head of the last. Add up the following two vectors, and label the resultant with the letter R.

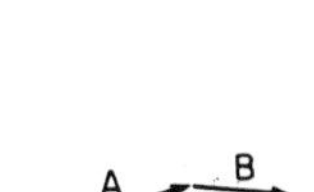

4. Vectors, like scalars, may be added in any order to obtain the same resultant. For instance,

3 + 3 + 7 = 7 + 3 + 3

Add the following group of vectors in at least three different orders and see if you always get the same resultant.

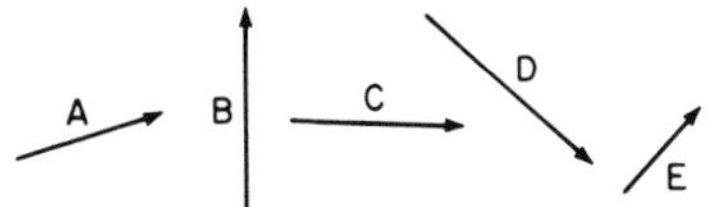

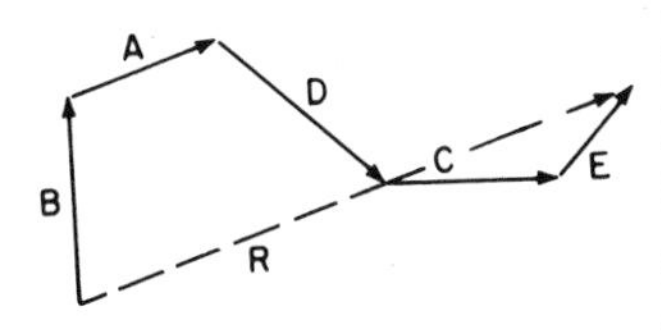

5. Another way to add vectors is to use the "Cartesian" method named after the French mathematician Descartes. A vector can be represented by placing the tail in the origin of a graph and giving the X and Y coordinates of its head. What are the X, Y coordinates of the following vector?

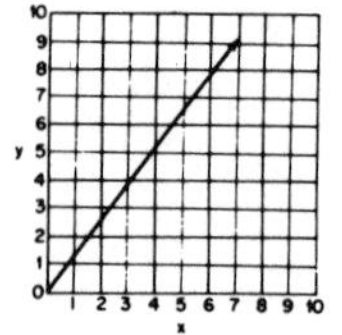

7,9

6. To add a number of different vectors, place them with their tails on the origin of a graph and add up their X and Y coordinates. Place the following vectors on a graph with their tails on the origin.

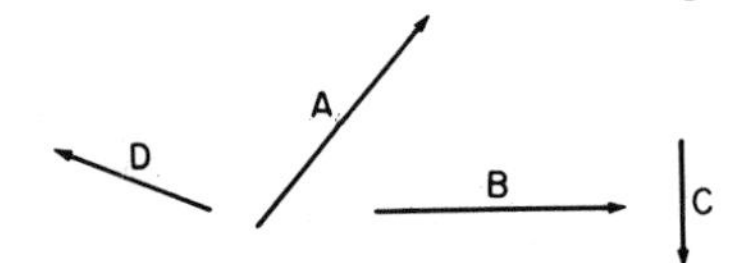

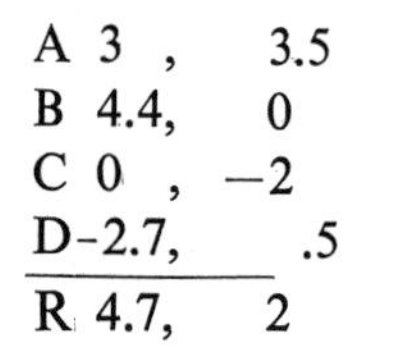

7. Find the X and Y coordinates of these vectors and add them up.

	X	Y
A	3 ,	3.5
B	4.4,	0
C	0 ,	−2
D	−2.7,	.5
R	4.7,	2

8. Draw the vector designated by the X and Y coordinates of the resultant.

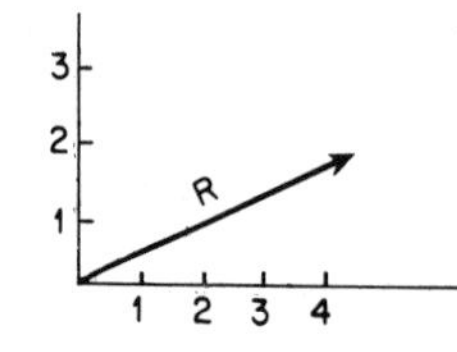

9. Add the A, B, C, and D vectors shown in question 6 by connecting tails to heads and drawing the resultant from the tail of the first to the head of the last. (Remember, the order in which they are added is unimportant).

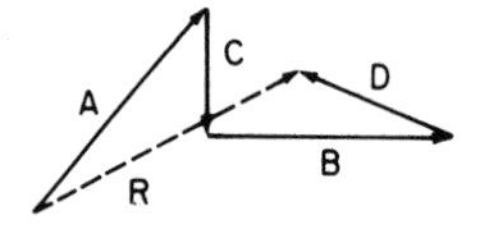

10. Are the resultants obtained by the Cartesian method and the head to tail method the same or different? Explain.

They are the same. They point the same direction and are of the same magnitudes.

11. Vectors may also be subtracted.
First consider the vector A.
If this A points in the positive direction, draw the vector which represents $-A$.

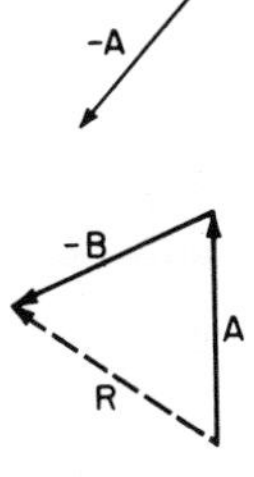

12. If $A - B = R$,
then $A + (-B) = R$. Subtract B from A

13. Another method of subtracting vectors may be represented as follows:

If $\vec{A} - \vec{B} = R$,

then $\vec{A} = \vec{B} + \vec{R}$

To subtract $\vec{B}$ from $\vec{A}$, join the tails of $\vec{A}$ and $\vec{B}$ and then join the heads of $\vec{B}$ and $\vec{A}$ with the resultant $\vec{R}$ so that $\vec{B} + \vec{R} = \vec{A}$. Subtract $\vec{B}$ from $\vec{A}$ by this method.

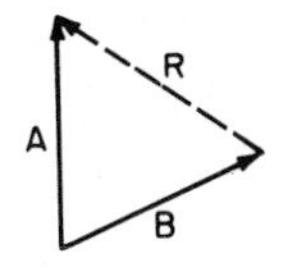

14. Are the resultants (R) obtained by these two methods the same or different? Explain.

They are the same, they point in the same direction and are of the same magnitude.

15. Vectors may also be subtracted by the Cartesian method.
Place the vectors A and B on a graph with their tails on the origin as in addition.

16. Find the X, Y components of A and B and subtract the B components from the A components.

$$\begin{array}{rr} 0, & 3 \\ -2.5, & -1 \\ \hline -2.5, & 2 \end{array}$$

17. Draw the vector R represented by these coordinates.

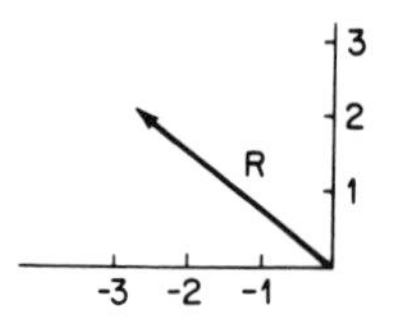

18. How does this compare with the vector R obtained by the previous two methods of subtraction.

same

19. Since we knew the lengths of the components of the vector R(−2.5, 2), we know its length is $\sqrt{(2.5)^2 + 2^2}$. Using the Pythagorean Theorem, find the length of vector R if its X component is -3 units and its Y component is 6 units.

$\sqrt{9 + 36} = \sqrt{45}$
$= 6.7$ units

20. Another way to find the length of R is by using trigonometry. If we call the angle between the axis and vector R, $\measuredangle\,\theta$ then $\tan\theta = Y/X = 6/3 = 2$. Since $\sin\theta = Y/R$, $R = Y/\sin\theta$. Find $\sin\theta$ when $\tan\theta = 2$ and solve for R.

A VECTOR LABORATORY

A useful laboratory exercise for enforcing the techniques of operating with vectors is the old force board experiment. However, I propose an innovation in this exercise which will help show some finer points of technique in such vector operations.

Suppose your force board was arranged as in Figure 5-1.

It is easy to add the three force vectors as in Figure 5-2 to show that the net force (the vectorial resultant of the three) adds to zero.

However, it is useful to ask your students to demonstrate that similar results can be obtained by resolving two of the forces into right angle components. Show that the two components which are perpendicular to the third force add up to zero, while the two components parallel to the third force add up to a net force that is equal and opposite to the third force. (See Figure 5-3.)

$$\sin\theta_1\cdot\vec{x} + \sin\theta_2\cdot\vec{y} = 0$$

$$\cos\theta_1\cdot\vec{x} + \cos\theta_2\cdot y + z = 0$$

Another type of problem which might be a useful exercise can be envisioned as follows:

An airplane is scheduled to fly from New York due west to Chicago, a distance of 690 air miles, in three hours. During the time of flight, he must contend with a wind from the north blowing with an average velocity of 30 miles per hour.

What velocity must the pilot maintain with respect to the air in order to land in Chicago on schedule?

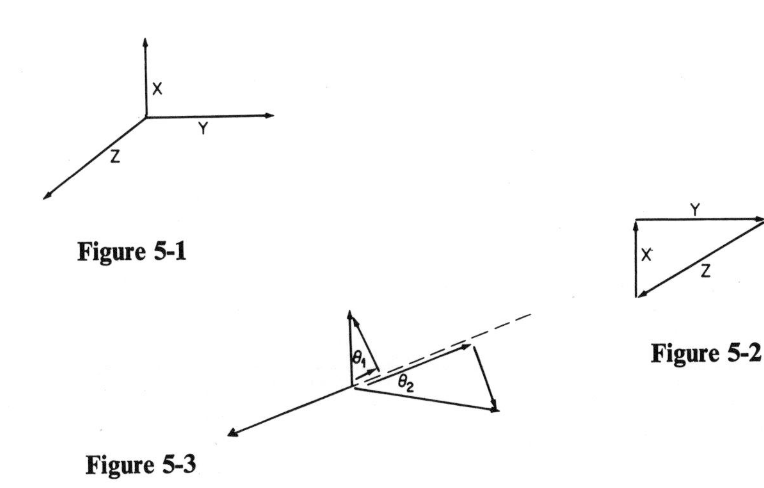

Figure 5-1

Figure 5-2

Figure 5-3

This is a navigation problem involving vectors. One can see that the desired velocity with respect to the ground must be 690 mi/3 hr. west = 230 mi./hr. west.

We wish then to have ($\vec{V}g$) 230 mi/hr west to be the resultant of the velocity of the airplane with respect to the air ($\vec{V}a$) plus the velocity of the air with respect to the ground (the wind velocity $\vec{V}w$).

$\vec{V}a + \vec{V}w = \vec{V}g$

$\vec{V}a = \vec{V}g - \vec{V}w$

But $Vg - Vw$ are known and may be represented by vectors as follows:

$Vg = 230$ mi/hr west
$Vw = 30$ mi/hr south
$-Vw = 30$ mi/hr north
(See Figure 5-6.)

We add $(Vg) + (-Vw)$ and find that the pilot must set his course at an angle north of west whose tangent is $30/230 = .13$, $\measuredangle = 7.4°$ and speed $= \sqrt{900} + \sqrt{53000} = 232$ mi/hr. His velocity with respect to air must be 232 mi/hr at a heading of 7.4° north of west.

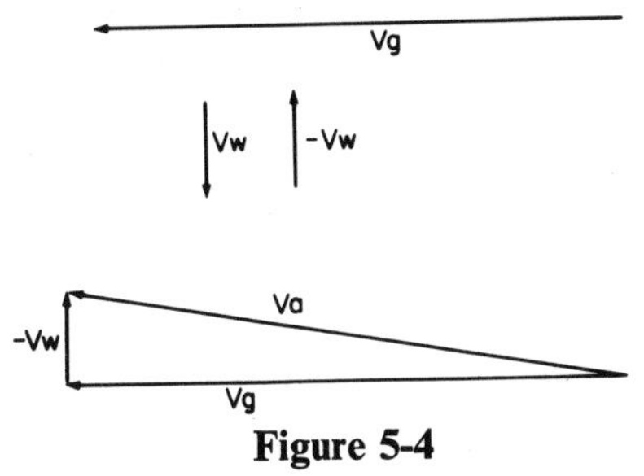

Figure 5-4

6

Kinematics, Rectilinear and Curvilinear

Concepts are not dogmatically introduced . . . but developed from facts. Theories and hypothesis come to discussion only when there is a need felt for them. Not through mentioning many dates and names, but through our genetic exposition does the historical moment gain its significance.

ERNST MACH

SOME ASPECTS OF KINEMATICS have already been introduced in Chapter 4 in order to illustrate the use of mathematical techniques. Indeed examples from kinematics are generally used by mathematics teachers for purposes of illustration and problem solving.

PARABOLIC TRAJECTORIES

A simple motion to study is the parabolic trajectory of a ballistic missile near the earth's surface. We have previously discussed accelerated motion when we derived $D = \frac{1}{2} AT^2$. We might now investigate motion having a constant velocity in one direction and acceleration in another direction.

The easiest trajectory to study is one in which the missile is launched from some height (y) with an initial velocity V_x parallel to the earth's surface.

The acceleration (g) due to gravity will then always be down (by convention, the negative direction) and (y) at any time (t) after the launching (when $t = 0$) will be $y = -\frac{1}{2} gt^2$. The distance moved in the X direction is simply $V_x t$. Since we will assume negligible friction and $X = V_x t$, solving for t ($t = X/V_x$) and substituting for t in the previous equation $y = -\frac{1}{2} g (X/V_x)^2$.

Since g and V_x are constant, the equation is of the form $y = -kx^2$ which should be recognized as the equation of a parabola.

Students can be asked to graph $y = kx^2$ to satisfy themselves that it will form the parabola predicted and then be shown pictorially by the diagram that follows.

First indicate constant velocity in the X-direction.

• • • • •

Then show acceleration in the Y-direction.

Finally, indicate both actions taking place at the same time.

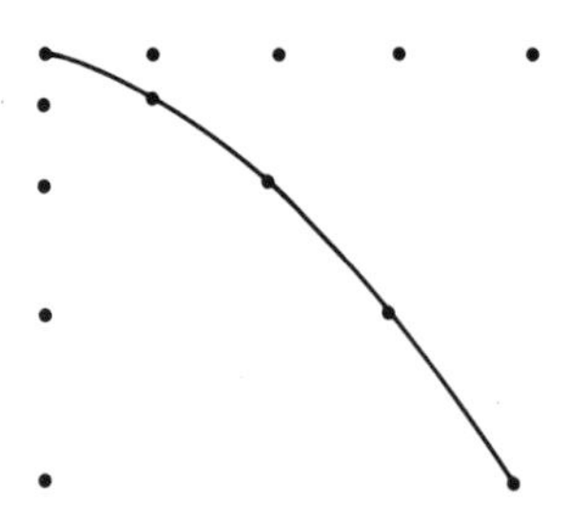

The parabolic path of such trajectories can be easily demonstrated by shooting a stream of water with some pressure out of a hose and observing the path the water takes. Strobed polaroid pictures of objects undergoing such motion would be useful for student scrutiny and analysis.

We can now move to the general case by suggesting that we launch an object at some initial velocity (V_0) at some angle θ such that V_0 is not parallel to, or perpendicular to, the earth's surface. The initial velocity V_0 can then be resolved into two component velocities V_y perpendicular to the earth and V_x parallel to the earth as in Figure 6-1:

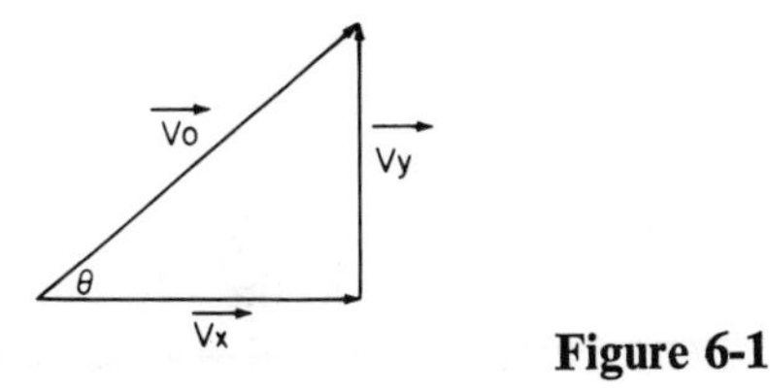

Figure 6-1

We conventionally assign the positive sign to the upward direction and the negative sign to the downward direction. The height (y), therefore, at any time (t) after launching can be described by the equation $y = V_y t - \frac{1}{2}gt^2$ where g is the gravitational acceleration. The range X would similarly be expressed by $X = V_x t$ since there is no acceleration parallel to the earth's surface. We can assume, if our missile is dense enough and V_0 is small compared to the terminal velocity, that frictional considerations will be very small. We can then obtain y in terms of x by solving for t in the second equation and substituting in the first.

$$t = \frac{X}{V_x}$$

$$y = \frac{V_y}{V_x} \cdot X - \tfrac{1}{2} g \frac{X^2}{V_x^2}$$

but $V_y = V_0 \sin\theta$ and $V_x = V_0 \cos\theta$,

$$\text{therefore } y = \frac{V_0 \sin\theta}{V_0 \cos\theta} \cdot X - \frac{g}{2} \frac{X^2}{(V_0 \cos\theta)^2} = \frac{\sin\theta}{\cos\theta} \cdot X - \frac{g}{2} \frac{X^2}{(V_0 \cos\theta)^2}$$

Now we can use this relationship to find out how far a projectile will go (its range) when launched at some X with velocity V_0. Y is set equal to zero (the projectile has gone up and come down again) and

$$\frac{\sin\theta}{\cos\theta} X = \frac{g}{2} \cdot \frac{X^2}{(V_0 \cos\theta)^2}$$

$$X = \frac{2\, V_0^2 \cos\theta \sin\theta}{g}$$

$$\text{and } X = \frac{V_0^2}{g} \sin 2\theta$$

The range of the projectile, therefore, depends on the angle of the launching and the initial velocity V_0. The maximum range will be achieved when $2 \cos \theta \sin \theta$ or $\sin 2\theta = 1$ or $\theta = 45°$.

It is also interesting to point out, or better yet lead your students to find, that the value for X is the same for any deviation from 45° on either side. That is X will be the same when $\theta = 50°$ or 40°, 60° or 30°, 75° or 15°, etc.

I have found it useful, when I find I have some students in my class who participate in track and field events like shot-put, to get them to predict the best angle at which to release the shot. Sometimes good discussions have evolved and a student usually volunteers that the best angle is slightly less than 45° in order to compensate for the height of the shot-putter.

Here a demonstration with a stream of water helps verify the predictions made by the calculations your students have down.

Many interesting problems such as time of flight, maximum height, etc., can be assigned and discussed.

"THE MONKEY AND THE HUNTER"

An interesting demonstration of the infallibility of the predicted trajectories can be accomplished if you wish to go to the trouble of constructing a "monkey and gun" apparatus.

Our equations predict that if a projectile is aimed at a target which begins to fall at the moment the projectile is launched, both will fall at exactly the same rate. If the (V_x) velocity of the projectile toward the target is sufficient so that they can collide before they both hit the ground, they will always collide.

For the gun, a spring laden device will do. The "monkey" would be an iron ball hung from an electro-magnet. A switch is placed in front of the gun so that it will turn off the electromagnet as soon as the projectile leaves the gun, and both the projectile and the ball can start falling at the same time. (See Figures 6-2A, 6-2B, 6-2C.)

The paths shown in Figure 6-2B are probably easily understood, though probably not immediately arrived at, without understanding the kinematics first, by hunters who are aware that to shoot a flying bird you must "lead it."

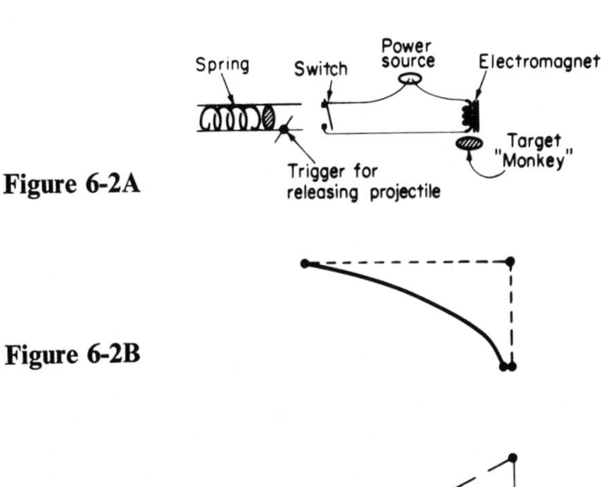

Figure 6-2A

Figure 6-2B

Figure 6-2C

The paths illustrated by Figure 6-C are not so obvious, yet it can be demonstrated that the "hunter" always hits his target regardless of the angle of his aim as long as he aims directly at the target and imparts sufficient velocity to his projectile so that it reaches the target before it hits the ground.

This can be justified as follows (see Figure 6-3):

$$Y_p = V_y t - \tfrac{1}{2} g t^2$$
$$Y_t = Y - \tfrac{1}{2} g t^2$$

For a collision, $Y_p = Y_t$

Therefore $V_y t - \tfrac{1}{2} g t^2 = Y - \tfrac{1}{2} g t^2$

$$V_y t = Y$$
$$V_x t = X$$

Eliminating t, $\dfrac{V_y}{V_x} = \dfrac{Y}{X}$

$$\tan\theta_p = \tan\theta_t$$
$$\theta_p = \theta_t$$

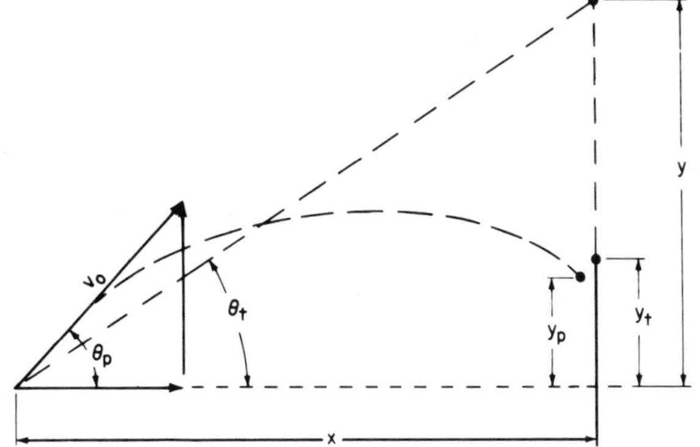

Figure 6-3

CIRCULAR MOTION

The motion of a body moving at constant speed in a circle is also of interest. Most students are already sure that they understand this motion. They ascribe it to some mysterious force called centrifugal force which causes a body moving in a circle to be forced out radially from the center. I would like to suggest a little demonstration to dispel this notion.

A Homemade Accelerometer

First build an accelerometer. An easy one to make involves an Erhlenmeyer flask, some string, a cork, and a rubber stopper.

The string is attached to the stopper and cork so that when the flask is filled with water the string is just short enough to permit the cork to float freely when the flask is held in the inverted position. (See Figure 6-4.)

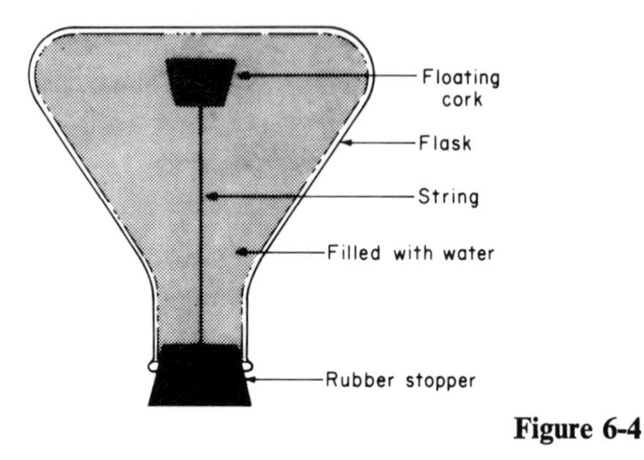

Figure 6-4

With this arrangement held firmly in your hand, let your students guess which way the cork will move when you suddenly push it forward. They will, of course, almost invariably suggest that the floating cork will move backwards.

This is not so. The cork is less dense than the water in which it floats and the greater inertia of the water will tend to have it resist the forward motion more than the cork will. The cork will move forward pointing in the direction of the acceleration.

It is interesting to digress a bit as we describe the accelerometer and suggest that this arrangement is also a very good demonstration of Albert Einstein's theory of general relativity. Einstein pointed out that if you were standing in a closed box and felt yourself attracted downward to the floor, there would be no experiment you could do inside the box to ascertain whether you were on a massive body with gravity pulling down or whether you were actually in space being accelerated upward. In other words, an acceleration in one direction is equivalent to a gravitational field in the opposite direction. When the accelerometer stands at rest or moves at a constant velocity, the cork points straight upward (opposite to the gravitational field downward). When the accelerometer is suddenly moved forward, the cork points forward as if there were a gravitational field backward, and when the device is slowed down (decelerated) the cork points backward as if its weight had shifted around.

When your students are convinced that the cork does really indicate the direction of the acceleration, then hold it out at arm's length and pivot in a circle. Call attention to the fact that as the accelerometer describes a circular path, the cork is pointing inward toward the center of the circle. Therefore, one might conclude that in circular motion the acceleration is inward.

Later in our study of dynamics, we will find that an acceleration must be caused by some force, and the acceleration must be in the direction of the force. Therefore, the force involved in circular motion is an inward force called a centripetal force, not that outward "centrifugal force" our students thought they knew all about.

A Geometrical Analysis

A more complete analysis might be demonstrated as follows:

If an object is moving at constant speed in a circular path, its instantaneous velocity always has the same magnitude, but since it is tangential to the path, it must always be pointing in a different direction as in Figure 6-5.

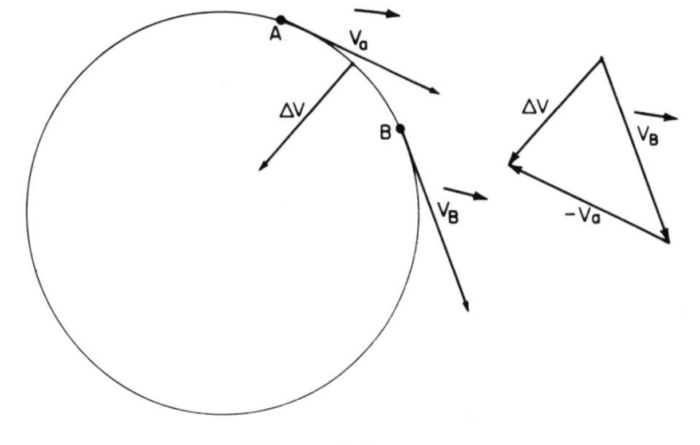

Figure 6-5

During the time (Δt) that the object moved from position A to position B, its velocity changed from $\vec{V}_a$ to $\vec{V}_b$ and the difference in velocity

($\Delta\vec{V}$) is obtained by subtracting $\vec{V}_a$ from $\vec{V}_b$ ($\Delta\vec{V} = \vec{V}_b - \vec{V}_a$) or $\vec{V} = \vec{V}_b + (-\vec{V}_a)$.

This is done vectorially. (See Figure 6-5.)

Note that when the vector $\Delta\vec{V}$ is placed with its tail on the path halfway between positions A and B (as would be appropriate since it denotes the average change in velocity between A and B), then $\Delta\vec{V}$ points directly to the center of the circle.

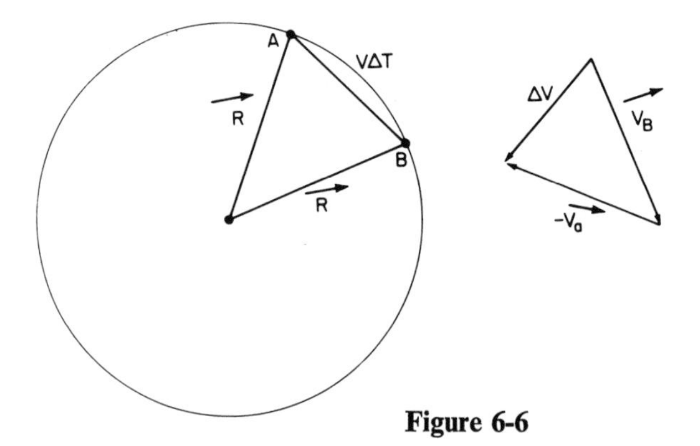

Figure 6-6

With Figure 6-6 in front of us, we can add the radial lines to A and B and the chord AB which is equal to $V\Delta T$ and use the geometry thus pictured to arrive at an expression for the centripetal accelerations in terms of the other parameters represented.

Note that the two triangles represented by R, R, $V\Delta T$ and V_b, $-V_a$, ΔV are similar since they are both isosceles triangles with mutually equal apex angles. The apex angles are proven equal since they are acute angles with mutually perpendicular sides. (See Figure 6-7.)

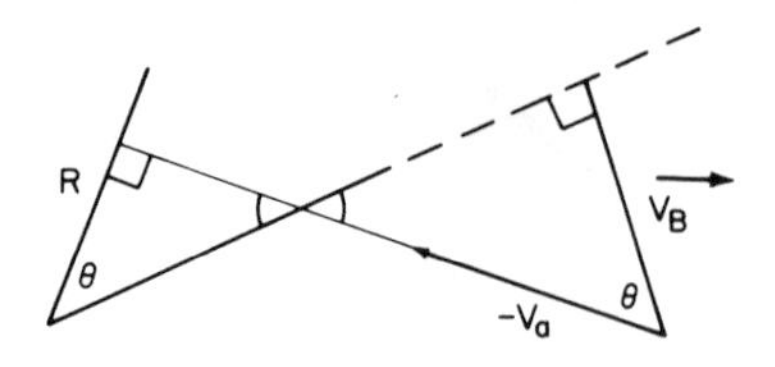

Figure 6-7

As these angles become small, the chord $V\Delta t$ approaches the actual path AB as a limit and the summation of all these very small paths adds up to the whole circle.

Using the similarity of the triangles, point out that corresponding parts of similar triangles bear the same ratio to each other and $\Delta V/V\Delta T$

$= V/R$ where we are interested now only in the magnitude of V and not its direction. Therefore $\Delta V/\Delta T = V^2/R$, but $\Delta V/\Delta T$, by definition, is the acceleration A and therefore $A = V^2/R$.

Since the speed V might also be denoted by $2\pi R/T$ (the circumference of the circle divided by the period in which the object circumscribes the complete circumference), therefore, $a = V^2/R = 4\pi^2R^2/(T^2/R) = 4\pi^2R/T^2$

This last is also a very useful expression for centripetal acceleration, as we will see later.

An interesting alternate derivation for centripetal acceleration is presented in the PSSC physics text.

As an object moves in a circular path, one might envision its displacement from the center of the circular path as a constantly changing radius vector. (See Figure 6-8.)

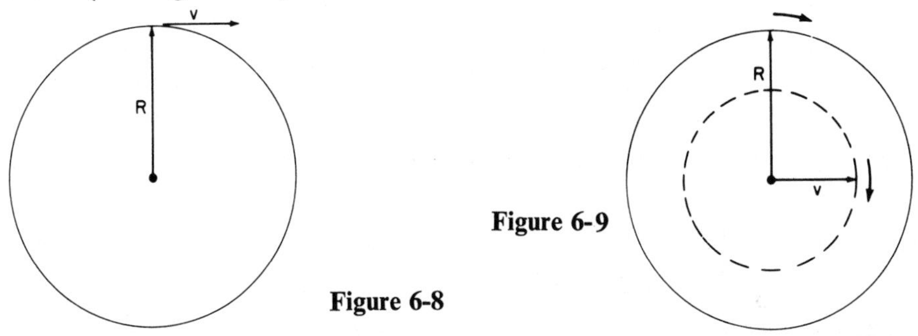

Figure 6-9

Figure 6-8

The change in displacement per unit time, in this case $\Delta R/\Delta T$, is the velocity V, and the magnitude of the velocity may be expressed as $V = 2\pi R/T$. The velocity V is also always changing. Since the instantaneous velocity at any time is parallel to the path, in this case perpendicular to the radius and, therefore, the end of the velocity vector is moving in a circle of radius V in velocity space, and as above, $\Delta V/\Delta T$ is the acceleration. (See Figure 9.) In this case,

$$A = \frac{2\pi V}{T} \text{ but } V = \frac{2\pi R}{T}, \text{ therefore } A = \frac{4\pi^2 R}{T^2}.$$

If we multiply this expression by R/R, we get

$$A = \frac{4\pi^2R^2}{T^2R} \text{ but } \frac{4\pi^2R^2}{T^2} = V^2, \text{ therefore } A = \frac{V^2}{R}.$$

A laboratory experiment which helps bear out this relationship requires very simple apparatus. The one suggested in the PSSC laboratory guide is quite good. I would, however, suggest that the students start off with a large force and gradually remove the weights supplying the centripetal force as they become more and more adept at manipulating the apparatus with very small forces (where it becomes more difficult).

It is necessary to determine the relationship between the force and the frequency of rotation (the reciprocal of the period). Since the force varies with the square of the frequency, graphing the force directly vs. the frequency would give a parabola. This, however, may not be apparent if the forces are large where the parabola begins to approach a straight line. When the student is encouraged to make trials with very small forces, the resulting graph begins to show a definite curve and will then encourage him to try graphing force against the square of the frequency.

The arguments in regard to centripetal vs. centrifugal force should be laid to rest very early in any discussion of circular motion. From the standpoint of a high school student who studies physics in inertial frames of reference where Newton's laws of motion hold, there is no centrifugal force. It is a name given to an apparent force because the observer has inadvertently transferred to an accelerated frame of reference.

Another form of motion which will be dealt with in more detail later is periodic motion, and specifically, simple harmonic motion.

In order to accustom my students to what is expected of them in the laboratory, one of the first homework assignments I give them is to find on what the frequency of a pendulum depends.

Assigning this type of laboratory for homework is a natural since the equipment involved is very simple (a weight on a string and a watch with a second hand). It also emphasizes a point I made earlier. A laboratory is any place where measurements are made.

Practically all my students report that varying the weight does not affect the frequency or period and that the amplitude of swing seems to have no appreciable effect. Almost all find that varying the length of the pendulum causes a change in the frequency, and some even discover an exponential relationship. I do not press too hard on quantitative results here, suggesting that we will investigate the pendulum more fully after we have studied dynamics and can discuss the force causing the pendulum's motion and the dependency of the period on the force. Most students, at this point, attain a sense of satisfaction at having designed an experiment, performed it independently, and arrived at some positive conclusions.

A sample problem and solution follows:

1. The moon travels in an orbit about the sun which is approximately 38×10^4 km in radius in a period of 27 days, seven hours and 43 minutes. What is the magnitude of its acceleration?

An analysis of the problem would indicate this to be a centripetal acceleration and

$$a = \frac{4\pi^2 R}{T^2} = \frac{4 \times (3.14)^2 \times 38 \times 10^7 \text{ meters}}{(27 \text{ days, } 7 \text{ hrs, } 43 \text{ min})^2}$$

$$= \frac{4 \times (3.14)^2 \times 38 \times 10^7 \text{ meters}}{\left(27 \text{ days} \times \frac{24 \text{ hrs}}{\text{days}} \times \frac{3600 \text{ sec}}{\text{hr}} + 7 \text{ hrs} \times \frac{3600 \text{ sec}}{\text{hr}} + 43 \text{ min} \times \frac{60 \text{ sec}}{\text{min}}\right)^2}$$

$$= 2.9 \times 10^{-3} \text{ meters/sec}^2$$

It would be interesting to check this acceleration at a distance of 38×10^7 meters from the earth's surface or $38 \times 10^7 + 6.4 \times 10^6$ meters from the center of the earth against the acceleration of 9.8 meters/sec² at the surface of the earth or 6.4×10^6 meters from the center of the earth.

$$\frac{g \text{ (the acceleration of a freely falling body at the earth's surface)}}{a \text{ (the acceleration of the distance of the moon from the earth)}} = \frac{R_m^2 \text{ (the square of the moon's distance from the center of the earth)}}{R_e^2 \text{ (the square of the radius of the earth)}}$$

$$\frac{9.8}{a} \text{ m/sec}^2 = \frac{(3.8 \times 10^8)^2 \text{ m}^2}{(6.4 \times 10^6)^2 \text{ m}^2}$$

$$a = \frac{41 \times 10^{12} \times 9.8}{15 \times 10^{16}} = 2.7 \times 10^{-3} \text{ m/sec}^2$$

The result previously obtained (2.9×10^{-3} m/sec²), when we calculated the centripetal acceleration of the moon, is a pretty fair approximation of this result.

7

Behavior of Light and Presentation of Geometric Optics

Science is built with facts just as a house is built with bricks, but a collection of facts cannot be called a science any more than a pile of bricks can be called a house.

HENRI POINCARE 1854—1912

THE ORDER IN WHICH WE TEACH the various subjects of physics is largely determined by the textbook we use and our own taste. It is important, however, that we take up the simpler subjects first so that what has been digested there can be used to understand more fully what comes later. I have usually followed the PSSC scheme of bringing in light and wave motion between kinematics and dynamics. It is a little easier for students and I find it useful to encourage students by some success before they tackle the more quantitative principles of dynamics.

Before working on the model for light, we usually look at the way light behaves in regard to reflection, refraction, and dispersion. A demonstration of the laws of reflection and some other interesting aspects of light behavior follows:

DEMONSTRATION OF LAWS OF REFLECTION

A piece of plate glass is set up on the demonstration table (I usually use one of my ripple tanks standing on an edge) with a candle in front of it and another identical candle sitting in the rear of it. The second candle must be positioned exactly on the image of the first. Identical bunsen burners or even small light bulbs can be used instead of candles.

All this is set up before the students come into the room. I then point out that I am going to demonstrate some properties of light and ask them to watch me carefully. The candle in the rear of the glass is lighted first and the students in front observe the flame through the glass plate. I then light the candle in front of the plate glass and while they observe both candles burning, an opinion I continue to nurture, I surreptitiously snuff out the flame of the candle in the rear. I do this as I point to the flame to ask if they can see it. The flame they continue to see on the rear candle is, of course, the image of the flame in front. Since the candles were originally positioned in each other's images, the effect is quite realistic. (See Figure 7-1.)

One then can do many things like seemingly permitting the flame to pass through your hand while saying "are you sure this candle is lighted?" Bring your eye down as if making a close examination so that to the students in front it will appear as if the flame is actually going through your head. When using light bulbs, one can make a great show of trying to put out the bulb in the rear. Then finally smash the bulb with a hammer and the bulb will still appear to be lit. It is very effective but somewhat more costly than candles or bunsen burner flames.

After you have amused the students sufficiently, remove the glass between the two candles and show them that the one in the rear was really not lit. If you do it right, there will be some surprise.

Your discussion with your students regarding the effect they had just witnessed should evoke the fact that the reason all saw the image of the flame in the same place was that the light hitting the glass surface was reflected according to a basic law of reflection. That is, the angle of reflection equals the angle of incidence. Another interesting property of light they had just witnessed was the fact that when light came to an interface (in this case, air and glass), even though the glass is a very good transmitter of light, part of the light was reflected at the interface. This is a very useful idea since you will be able to draw on it later in discussing whether or not light appears to behave more like a particle or a wave.

Laboratory Exercise on Laws of Reflection

The laboratory exercise which might have preceeded this demonstration

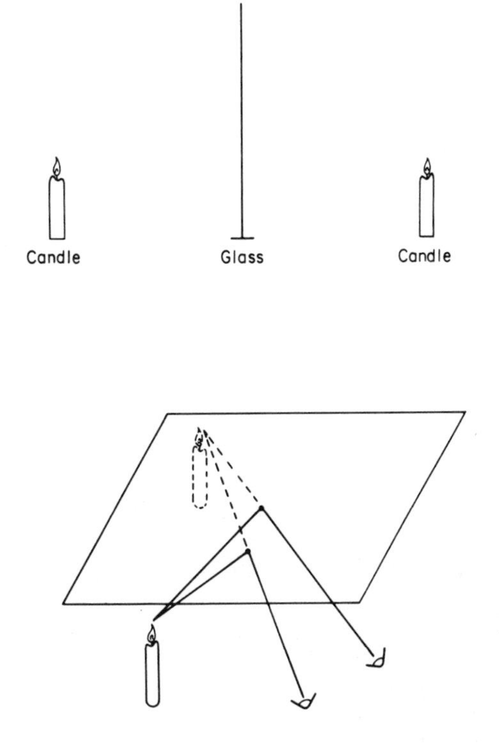

Figure 7-1

is the usual one in which students find the location of the image of a pin in a plane mirror. I strongly recommend that they be asked to find the location of the image by two methods.

Parallax Method of Image Finding

The parallax method of finding an image, by lining up the image viewed in the mirror with a movable object seen over the top of the mirror and moving this object until the image and object appear to be in the same place no matter what the viewing angle is, is very easy to do. Besides confirming the location of the image in a plane mirror, it will help students develop the technique of finding an image by parallax. They will be able to use this technique in situations where the use of this technique is more difficult, and yet may be the only way to locate experimentally the position of the image.

Finding an image in a mirror by pinsighting is probably more meaningful for developing the laws of reflection. For instance, after the student has lined up his pins with the image of another pin as seen in the mirror, he can draw the necessary ray diagram to show by his own measurement with a protractor that the angle of reflection is always equal to the angle of incidence. (See Figure 7-2.)

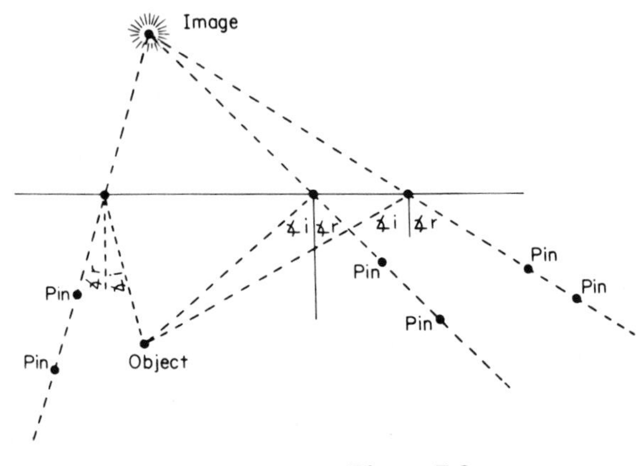

Figure 7-2

DERIVATION OF SNELL'S LAW OF REFRACTION

The laws of refraction are also easy to demonstrate and are preferably discovered by your students (with your help) in the laboratory.

The semi-circular, transparent, thin-walled cheese boxes used in the PSSC laboratory exercise on Snell's law make an excellent tool for this exercise.

Supplementary Refraction Experiment

One might supplement this laboratory with a similar one in which students find the rays by pinsighting as the light goes in and out of a transparent glass or plastic block with parallel sides.

The purpose of this laboratory exercise is to reinforce the notion that a ray of light entering a medium with parallel sides will emerge from the other side parallel to the path of entry, though possibly displaced. (See Figure 7-3.)

This can be predicted as follows: Snell's Law states that the ratio of the semichords in the two media is always the same. This ratio is called the index of refraction between the two media. (See Figure 7-4.)

Of course this is more familiarly stated as the ratio of sin $\sphericalangle i$ to sin $\sphericalangle r$.

Thus

$$n_{\text{air to water}} = \frac{\sin \sphericalangle \text{ in air}}{\sin \sphericalangle \text{ in water}}$$

The converse must also be true.

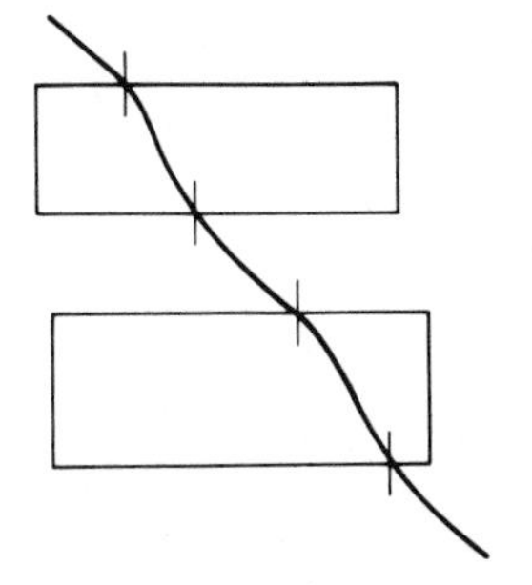

Figure 7-3

$$n_{\text{water to air}} = \frac{\sin \sphericalangle \text{ in water}}{\sin \sphericalangle \text{ in air}}$$

An algebraic solution to predict the experimental evidence might run something like this:

$$n_{\text{air to glass}} = \frac{\sin \sphericalangle \text{ air}}{\sin \sphericalangle \text{ glass}}$$

$$n_{\text{glass to air}} = \frac{\sin \sphericalangle \text{ glass}}{\sin \sphericalangle \text{ air}}$$

$N_{a \to g} \times N_{g \to a} = 1$ (no deviation)

Let us do the same thing for a more complicated situation. Suppose light penetrated a sheet of glass with parallel sides, then a layer of water, and finally went through a layer of kerosene.

$$N_{a \to g} = \frac{\sin \sphericalangle a}{\sin \sphericalangle g}$$

$$N_{g \to w} = \frac{\sin \sphericalangle g}{\sin \sphericalangle w}$$

$$N_{w \to k} = \frac{\sin \sphericalangle w}{\sin \sphericalangle k}$$

$$N_{k \to a} = \frac{\sin \sphericalangle k}{\sin \sphericalangle a}$$

$$N_{a \to g} \times N_{g \to w} = \frac{\sin \sphericalangle a}{\sin \sphericalangle g} \times \frac{\sin \sphericalangle g}{\sin \sphericalangle w} = \frac{\sin \sphericalangle a}{\sin \sphericalangle w} = N_{a \to w}$$

$$N_{a \to w} \times N_{w \to k} = \frac{\sin \sphericalangle a}{\sin \sphericalangle w} \times \frac{\sin \sphericalangle w}{\sin \sphericalangle k} = \frac{\sin \sphericalangle a}{\sin \sphericalangle k} = N_{a \to k}$$

$$N_{a \to k} \times N_{k \to a} = 1$$

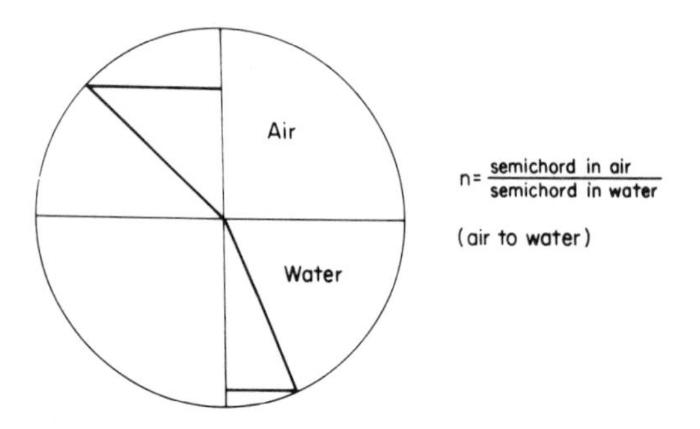

Figure 7-4

The ray, therefore, comes out of the kerosene surface into air parallel to the path in which it entered the bottom glass surface from the air.

CRITICAL ANGLE AND TOTAL INTERNAL REFLECTION

One of the difficult concepts for beginning students to grasp is the critical angle and exactly what it signifies.

The critical angle can very easily be demonstrated by shining a pencil of light through a semi-circular glass blank at the flat glass to air interface. As you gradually increase the angle of incidence at which the light comes in contact with the interface, the refracted ray in the air becomes less and less perceptible while the reflected ray in the glass becomes more and more pronounced until the angle of refraction in the air becomes 90°. At that point, no light escapes the surface. It all returns along the strongly accentuated reflected ray.

Part of the confusion, I believe, is due to the convention of expressing the index of refraction of water as $n = \sin \measuredangle a/\sin \measuredangle w$, and deriving therefrom the expression for the critical angle in terms of the index of refraction.

$$n = \frac{\sin \measuredangle 90}{\sin \measuredangle c} = \frac{1}{\sin \measuredangle c}; \text{ therefore, } \sin \measuredangle c = \frac{1}{n}$$

Some students are thus led to believe that the incident light must be in the air since the index of refraction had previously been defined as $n = \sin \measuredangle i/\sin \measuredangle r$.

It must be pointed out that total internal reflection occurs only when light coming through the denser medium strikes an interface with a less dense medium (one in which it will be refracted away from the normal) at the critical angle or greater. The critical angle should be defined as the angle of incidence in the more dense medium when the refracted angle in the less dense medium becomes 90°.

Determining Critical Angle by Experiment

A laboratory variation of the exercise for determining the index of refraction can be used for clarifying exactly what is meant by the critical angle and total internal reflection. A transparent circular cheese box can be filled with water and students asked to determine the critical angle of water to air as follows (see Figure 7-5):

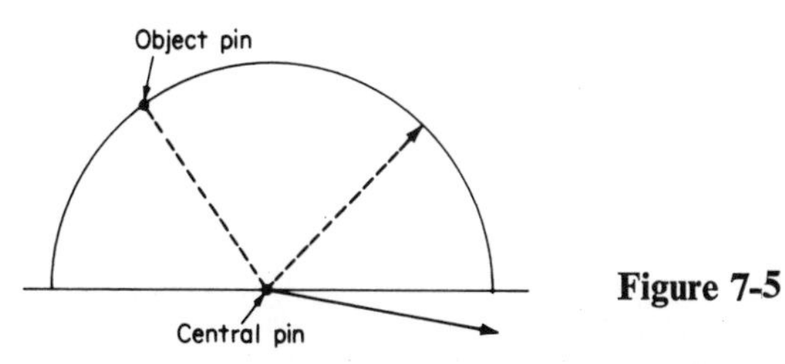

Figure 7-5

Pins are moved around the curved wall of the water container and lined up with a pin in the center of the straight wall of the container. The incident angle in the water can be increased by moving the pin around the curved wall. Sightings can then be made with two other pins. One will determine the position of the reflected ray by sighting outside the curved wall side on the central pin and the image of the object pin. The other pin determines the path of the refracted ray in the air by lining up with the object and the central pin. The object pin is moved until the sighting pin on the flat side indicates a 90° refracting angle and the object can no longer be seen. It will be quite evident that the image of the object from the sighting pin on the curved side will be very distinct. The incident angle in the water can now be measured. This is the critical angle. The reciprocal of the sine of this angle should be equal to the index of refraction for water previously determined.

Dispersion

The demonstration of the dispersion of white light by a medium with nonparallel opposite sides (a prism) is well known and need not be discussed here. The important thing that must be emphasized here is that such a dispersion would not be possible unless the index of refraction of this media for the various color components of white light differs slightly.

LENSES AND MIRRORS

The investigation of the properties of reflection of curved surfaces starting with combinations of plane surfaces and the refractive properties of lenses and working with combinations of prisms are useful ways of introducing the subject.

If there is to be any clear understanding of geometrical optics at this level, students MUST be drilled in drawing ray diagrams. I generally assign, for homework, the ray diagrams for the following cases:

1. When the object is more than two focal lengths from the mirror or lens.
2. When the object is exactly two focal lengths away.
3. When the object is between one and two focal lengths.
4. When the object is on the focal point.
5. When the object is less than a focal length from the mirror or lens.

The above, of course, is assigned for converging lenses and mirrors.

For diverging lenses and mirrors, I suggest that they draw enough ray diagrams to come up with some common rule.

Lens Equations

From the ray diagrams, one can easily derive a very useful form of the lens equation.

Students should be encouraged to make this derivation from any of the ray diagrams made for mirrors and lenses. (See Figure 7-6.)

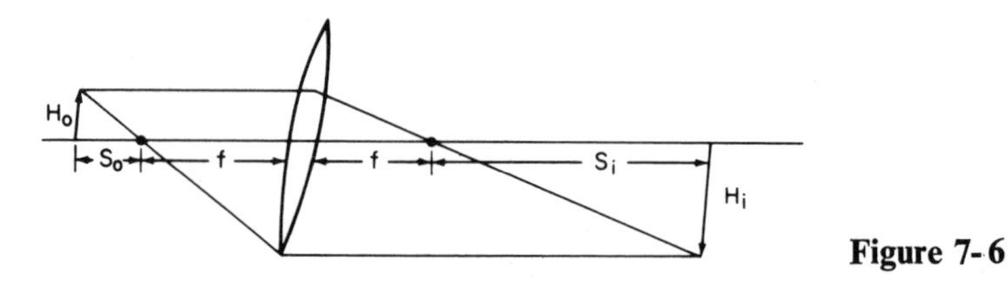

Figure 7-6

By moving H_1 and H_0 between parallel rays to make similar triangles and by using the similar triangles to show that $H_i/H_0 = f/S_0 = S_i/f$, we arrive at $f^2 = S_i\,S_0$.

This is a very useful equation, although most people are familiar with the more conventional expression,

$$\frac{1}{f} = \frac{1}{D_0} + \frac{1}{D_i}.$$

This can be derived from the less familiar expression as follows:

$D_0 = S_0 + f$ and $D_i = S_i + f$

substituting in $f^2 = S_iS_0$

$f^2 = (D_1 - f)\,(D_0 - f)$

$f^2 = D_1D_0 - fD_0 - fD_1 + f^2$

$D_1D_0 = fD_0 + fD_1$

$$\frac{1}{f} = \frac{D_0 + D_1}{D_1D_0} = \frac{1}{D_0} + \frac{1}{D_i}$$

Generations of students have had problems deciding when D_1 or D_0

was positive or negative. Many have problems handling reciprocals, and a casual inspection of an equation like $1/f = 1/D_1 + 1/D_0$ does not reveal much. $f^2 = S_1S_0$ says much more. If S_0 is negative, S_i must also be negative since f^2 must be positive (since f cannot be the square root of a negative number).

A good convention for sign follows:

For converging lenses and mirrors f is positive; for diverging lenses and mirrors f is negative.

For converging lenses and mirrors, Si and S_0 are both positive when a real image is formed. S_i and S_0 are both negative when a virtual image is formed. For diverging lenses and mirrors S_i and S_0 must always be positive even though only virtual images are formed.

Let us see what other information we can glean from inspection of $f^2 = S_iS_0$.

One can see that when S_0 equals f, S_i must also equal f. This predicts that when an object is placed two focal lengths from a converging lens, the image will be formed exactly two focal lengths behind the lens. For a converging mirror, the analogous case is predicted—the image will be superimposed on the object if the object is exactly two focal lengths from the mirror.

It is also readily perceivable that as S_0 gets smaller, S_i gets larger and when S_0 is infinitesimally small, S_i is infinitely large. That is, when S_0 is on the focal point, all rays emanating from the object are parallel to each other. The opposite is also true as seen from the equation. As S_0 approaches infinity, S_i approaches zero since the product must be a given real number. This predicts that all parallel rays coming from an object infinitely far away must go through the focal point.

Another problem which sometimes occurs is that D_i and D_0 are measured from the optical center of a lens. When very thin lenses are used, it is not much of a problem. But when a thick lens is used, especially one that is not symetrical, the optical center must first be determined. S_i and S_0 are measured from the focal points. These are definite points, and thus the thickness of a lens poses no problem.

If time dictates choosing between optics laboratories of mirrors or lenses, it should be strongly recommended that lenses offer students much more satisfaction. It is a lot harder to locate a real image of a mirror when the object keeps getting in the way than it is of a lens since the object is now on the other side.

It is at this point that skills gained by parallax location of images in a plane mirror come in handy. Finding the virtual image of a lens or curved mirrors is only possible by parallax.

Actually, once the student has drawn the ray diagrams for mirrors and lenses, seen some of your demonstrations, and participated in some of your discussions, no real purpose will be served in doing both laboratories since the geometry is so similar.

Finding S_i and S_0 for a lens (graphing S_0 vs $1/S_i$ and getting a straight line whose slope is f^2) and verifying f by finding the focal point experimentally by focusing light from an infinitely distant object, is most satisfying. This gives students a real sense of participation in a laboratory experiment. It is an easy lab and well worthwhile. If nothing else, it promises success to so many students who are otherwise so often frustrated.

Optical Problems

The assignment and discussion of optical problems help cement the concepts introduced here.

One such problem which I find useful for discussion is a problem in one of the Home, Desk and Laboratory sections of the first issue of the PSSC text involving the calculation of the magnification of a compound microscope. It is of interest to physics students since they have all previously used microscopes in biology. It was then a "black box" to them, and it serves as a very good illustration of the usefulness of all the relationships previously derived.

It would not be out of place to review the problem and its solution here.

"What is the magnification of a compound microscope whose ocular has a focal length of 2 cm and whose objective lens has a focal length of 4 mm?"

The two lenses are separated by a barrel that is 22.3 cm long. It is first useful to discuss how a compound microscope works. Since $H_i/H_0 = f/S_0$, it is necessary to bring the objective very close to the object being examined in order to magnify it. That is, S_0 must be very small. This will make S_i very large and the image will be formed high up in the barrel of the microscope. This real image now becomes the object for the ocular lens. It must be thrown inside the focal point of the ocular lens since the image of the original object is still upside down as was the real image made by the objective lens. Therefore, the image made by the ocular lens is a virtual image and has not been reinverted. Furthermore, since the viewer's eye is almost in contact with the ocular when the final image is brought into focus, it must have been formed back down the barrel on the same side of the ocular lens as its object is. Since the viewer focuses until the final image is approximately 25 cm from his eye (the average comfortable viewing distance for normal eyes; it differs slightly among individuals, and therefore the magnification of the microscope will differ slightly, depending on who is looking through it), we will use 25 cm as the distance of the image from the ocular (D_i).

It would be best here to draw our diagram schematically and refer to it in our ensuing discussion of the problem. (See Figure 7-7.)

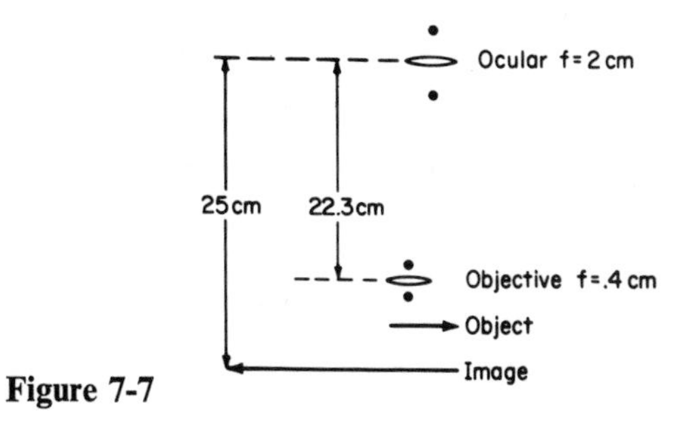

Figure 7-7

We know the distance of the image from the ocular and, therefore, it is easiest to start with the magnification of the ocular.

$S_i = 27$ (25 cm from lens + 2 cm from image focal point)

$$\frac{H_i}{H_0} = \frac{S_i}{f} = \frac{27}{2} = 13.5$$

We can now precisely locate the position of the real image, since

$$f^2 = S_i S_0,\ S_0 = \frac{f^2}{S_i} = \frac{4\text{ cm}}{27}.$$

The image is 4/27 cm inside the lower focal point of the ocular and is, therefore, $2 - \frac{4}{27}$ or $1\frac{23}{27}$ cm from the ocular. Since the ocular is 22.3 cm from the objective, or 21.9 cm from the objective focal point, S_i for the objective is $21.9 - 1\frac{23}{27}$ (approximately $21.9 - 1.9$), or 20.0 cm.

Magnification for the objective then is:

$$\frac{H_i}{H_0} = \frac{S_i}{f} = \frac{20.0}{.4} = 50.$$

The magnification of the microscope is the product of the magnification of its two lenses, $13.5 \times 50 = 675$.

8

Behavior of Waves in Mono-Dimensional Media

AFTER INVESTIGATING the behavior of light, it is appropriate to ask what light is. After some discussion of a particle model for light and finding it wanting (to be discussed more fully later), we are forced to look for another model. We start to investigate the behavior of waves to see if it matches what we have already discovered about light.

The group of ripple tank experiments outlined in the PSSC laboratory guide are excellent teaching laboratories. With proper guidance, students can spend every day, for about two weeks, in the laboratory and emerge with some real understanding of wave behavior.

One important point must be made here. While ostensibly the motivation for studying waves is to get to a better understanding of the nature of light, it is necessary, and this is your real motivation, to study waves as a method of energy transfer in general. Students with this background will be better equipped for further study of radio transmission, sound, gamma radiation from unstable nuclei, etc.

Students are first introduced to waves via mono-dimensional media like ropes, slinkies, wave machines, etc.

WAVE MACHINES

An effective wave machine can be made by soldering welding rods to a long steel wire. This can be hung from the ceiling in the classroom as shown in Figure 8-1.

A torsion wave started up the spine of this device will result in easily seen, slow moving transverse waves along the edge of the ribs.

A satisfactory machine is one designed for wave demonstration at the Bell Telephone Laboratories. This machine is distributed by Allegri Corporation of Nutley, New Jersey.

Another almost as satisfactory wave machine can be constructed by stretching wire across the front of the classroom and suspending beneath from about $2\frac{1}{2}$ to 3 feet of string, hung on the wire with paper clips, a slinky, which can be stretched across the room or slid down the wire to one side of the room to be secured when it is not in use. (See Figure 8-2.)

The slinky should be supported at every second or third coil to keep it from sagging when stretched out for demonstrations. One of the advantages of this arrangement over the ones previously described is that one can demonstrate both longitudinal and transverse waves, whereas the Bell Telephone machine, a ribbed spine device, and the other ribbed spine hung from the ceiling are very good, but only for transverse waves.

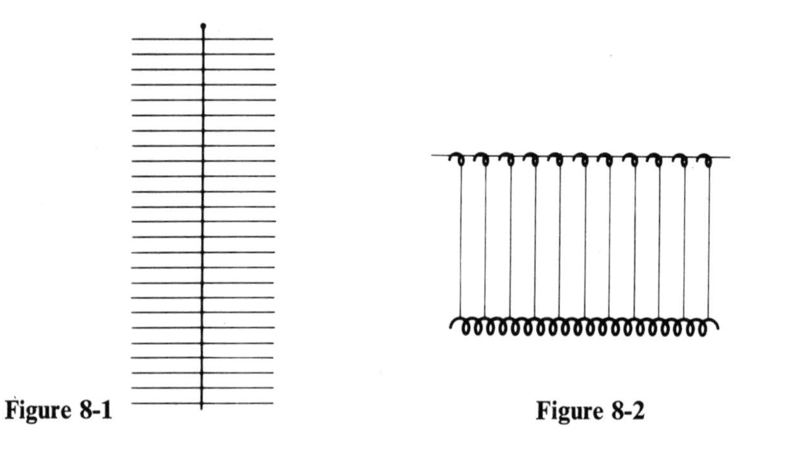

Figure 8-1 **Figure 8-2**

ANALYSIS OF PULSES

A pulse moving slowly along the proper mono-dimensional media can be analyzed and studied carefully. First it is obvious that as the pulse moves along the medium, the medium is displaced perpendicularly to the propagation of the pulse, and what is moving is only the tendency to be displaced from rest and back to rest.

If your students would graph the vector displacement of a point of the medium against time, they would arrive at the shape of the pulse going through the medium. (See Figure 8-3.)

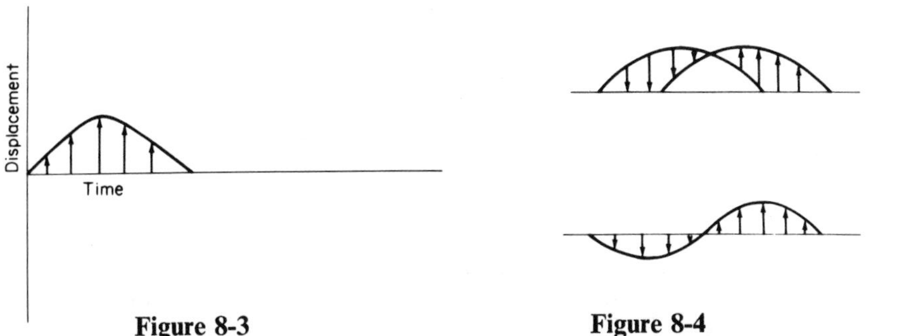

Figure 8-3

Figure 8-4

It would also be interesting to graph the velocity of the medium at any instant against the length of the pulse. This is done by graphing the change in displacement per unit time against the length of the pulse. $\Delta\vec{D}$ per unit time can be arrived at by inspection of the displacement vs. time graph and visualizing the pulse moving forward a short distance. (See Figure 8-4.)

It then becomes obvious that the leading edge of the pulse is moving up while the trailing edge of the pulse is moving down. One can also point out that the leading and trailing edges, while moving in opposite directions, are moving faster than the center of the pulse. (Incidentally, this kind of analysis is only possible if students have studied vectors before tackling wave motion.)

VELOCITY OF PULSES

The easiest thing to measure is the velocity of the pulse. This can be done with a stop watch or by counting slowly as the pulse moves from one end of the medium to the other. It should soon become evident, as students try to send pulses of various shapes through the media, that the velocity of the pulse is constant and does not depend on the size or shape of the pulse.

The medium is then varied by using a different ribbed spine (provided in the Bell Telephone—Allegri machine) or, if you are using a slinky, by changing the tension on the slinky. The point taught should be obvious to all—*the velocity of a pulse depends only on the medium through which it is propogated.*

TERMINOLOGY AND RELATIONSHIPS

One then sends a series of periodic pulses through the medium in a regular wave train. At this point you ought to define what is meant by wave-length, amplitude, frequency, crest, and trough. (See Figure 8-5.)

Show that by increasing the frequency you cause the wave length to become shorter, and vice versa. The argument for this effect might be presented as follows:

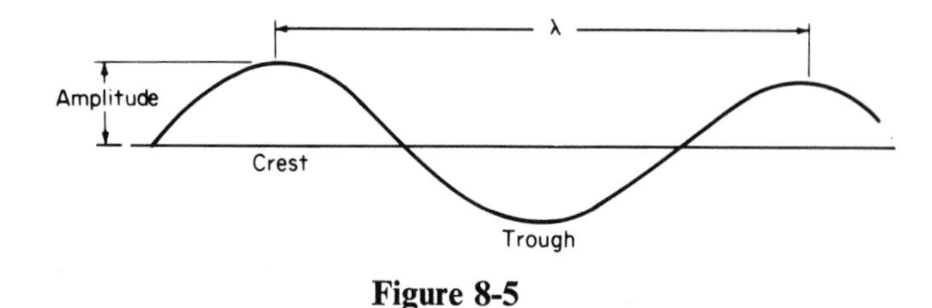

Figure 8-5

You have already shown that the velocity of the wave pulse depends only on the medium. Suppose this velocity was one meter per second and you cause 10 pulses per second to be propagated into the medium. Since the first pulse has advanced only one meter, the other nine pulses must be crowded into the one meter space behind them, so the space between each pulse is one tenth of a meter. We then arrive at an equation $V = \lambda\nu$; that is, the velocity of propagation is equal to the frequency times the wave length (1 meter/sec = 10/sec × .1 meter).

Here one must be sure to point out that while velocity is dependent on the medium, the frequency depends only on the source of the disturbance, and the wave length, therefore, depends on both the medium and the source. Thus, a train of waves of given frequencies will have different wave lengths in different media.

POLARIZATION OF WAVES

While you are propagating waves on your wave machine, the concept of polarization might be introduced. As a transverse wave moves along the medium, it is easy to point out that but for the limitation of the particular machine you are using, it would be possible for the medium to be disturbed in an infinite number of planes perpendicular to the direction of propagation. The hanging ribbed spine previously mentioned is ideal for demonstrating this.

If you are using the hanging slinky, a longitudinal wave can be propagated, and it is obvious with no prompting on your part that polarization is not a parameter for longitudinal waves.

The relationship between energy and amplitude ought also to be demonstrated. If you vibrate the medium harder (do more work on it) while maintaining the same frequency, you change the amplitude of the pulses. If you maintain the same amplitude, but increase the frequency, you are also doing more work per unit time.

This might be a useful idea to plant for resurrection when it is time to discuss quantum and the relationship between energy and frequency.

REFLECTION OF PULSES

The fact that pulses are reflected when they reach a barrier (a place where the energy cannot continue in the direction of propagation) is predicted by the law of conservation of energy. That it is reflected right side up when the barrier is a free end can be experimentally demonstrated and argued as follows:

If the end of the medium is free to move up and down, nothing is changed except that the medium is discontinued and the pulse must reverse its direction in order to conserve energy.

If the end of the medium is rigidly bound, the part of the medium just before the bound edge cannot transfer its energy in the direction of propagation, so it just continues to move below the rest point to a negative displacement equal in amplitude to its previous positive amplitude. Each succeeding part of the pulse repeats the process as the leading edge of the pulse proceeds back along the medium in the direction from which it came.

It is then a natural continuation of this line of reasoning to suggest and demonstrate that if the discontinuity is a juncture with another medium, then some of the energy travels across but some of it is always reflected, since, if the new medium is one in which the pulse travels faster (i.e., the wave length gets longer), it approximates a free end (that is, freer than if the medium were continuous), and the reflected pulse comes back right side up. If, on the other hand, the pulse crosses an interface wherein the new medium is one in which it travels more slowly (the wave length gets smaller), the reflected portion of the wave must come back upside down.

Students must, of course, be given the opportunity of working with these wave machines independently during a laboratory period. Some may glean all the information necessary for understanding wave phenomena from these independent laboratories, but realistically, we all teach mixed populations, and careful supplementary demonstrations and arguments by the teacher are necessary for all to understand. One may use the quicker students to help argue with those whose understanding comes a little more slowly.

Having successfully shown the effect of interfaces, one can argue then that the amount of transmission and reflection then depends only on how far apart are the two media in the velocity with which they permit energy to be propagated.

A neat demonstration can be made here for "impedance matching," as is done by Dr. Shive of Bell Telephone Labs when he demonstrates wave phenomena with the Bell Telephone Wave Machine. If the interface is a gradual one, reflection is minimized and transmission is increased.

This can be a problem in power transmission when electric waves (alternating currents) are transmitted along power lines. Junctions between dissimilar carriers are fitted with a device that makes for less of a sharp interface in order for a maximum amount of energy to be transmitted across the interface.

In optics, the analogy might be the case in which lenses are coated with a transparent substance whose index of refraction lies between glass and air so that more light is transmitted through the glass and less is refracted.

In acoustics, the use of a megaphone by a cheerleader might be observed. One must use a great deal of energy to be heard above a roaring crowd if the sound generated in the larynx must escape from the confines of the buccal cavity suddenly into open space. If the interface between the buccal cavity and the open air is made more gradual through the use of a megaphone, less sound is reflected back into the mouth, and the cheerleader need not work so hard to make herself heard.

INTERFERENCE

Interference phenomena can be easily studied and understood by means of demonstrations and experiments with your wave machines.

Two pulses started at opposite ends will interfere constructively as they go through each other, and the resultant interference can be analyzed by addition of the vector displacements. (See Figure 8-6.)

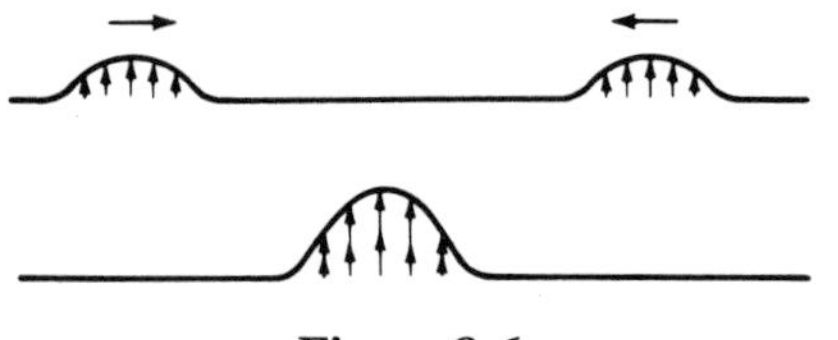

Figure 8-6

Another interesting fact to observe and point out is that once the pulses go through each other, there is no memory of the passage. Each pulse looks alike and behaves the same as if there never was another pulse to interfere with.

Two similar pulses started at opposite ends and 180° out of phase with each other will show destructive interference and can also be justified by vector addition of displacements. (See Figure 8-7.)

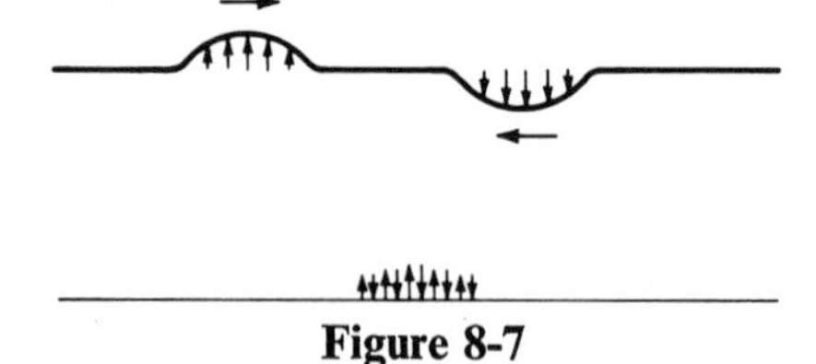

Figure 8-7

An interesting question to ask is why, at the instant of complete destructive interference when the medium appears to be momentarily at rest, do the pulses continue on? The answer is that, while there is apparently no potential energy stored in the distortion of the medium the medium is moving fastest then—the energy is all kinetic. This can easily be ascertained by adding up the velocity vectors. (See Figure 8-8.)

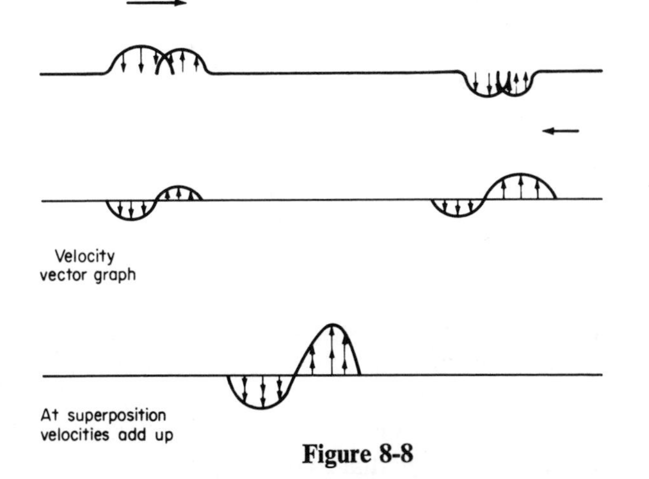

Figure 8-8

In the case of the two crests constructively interfering, the displacement vectors add up to make a larger displacement and the velocity vectors cancel out. This is because the leading edge of one pulse and the trailing edge of the other pulse with which it superimposes, when there is complete interference, are moving in opposite directions perpendicular to their velocity of propagation in the medium. (See Figure 8-9.)

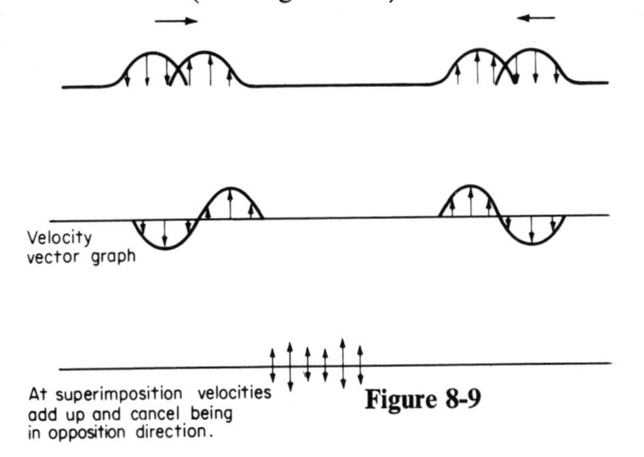

Figure 8-9

This would indicate that while the combined pulse is most distorted at that time, having maximum potential energy, it is actually momentarily at rest with zero kinetic energy. One could predict, therefore, that if the medium were merely held distorted, in the shape it assumes at the point of complete constructive interference, and then were suddenly released, two pulses would leave resembling the two pulses which caused the original constructive interference. This can then be demonstrated for verification.

STANDING WAVES

A series of pulses can then be generated down the medium with one end of the medium fixed. Standing waves will then be observed at certain frequencies. Some discussion can then ensue as to why certain frequencies produce standing waves and others do not.

The explanation should first point out that since the velocity in the medium is constant, the wave length imposed on the medium is a result of the frequency. A standing wave can only exist when the length of the medium can be divided into a whole number of half wave lengths. This occurs because a nodal point must exist at each end. Of course, the nodal point near the end where the disturbance is, is slightly away from the end since the oscillating point cannot be a nodal point. (See Figure 8-10.)

You can also demonstrate that as the frequency is increased, the standing waves get shorter. This reinforces the concept that the product of the wave length times the frequency is equal to a constant; i.e., the velocity of the wave in the medium.

Some digression might be made here regarding the physics of stringed instruments. The violinist tunes his strings by changing their tension—thus changing the velocity of the wave traveling along the string. When the string is bowed or plucked, the fundamental wave length of the standing wave must be twice the length of the musical string, the distance between two nodes. The frequency of vibration and the resulting frequency of the emitted sound is, therefore, the velocity with which a wave travels down the string divided by twice the length of the string. Violinists, as do players of most stringed instruments, change the frequency of the emitted sound by fingering the string. They thus change the position of a node and change the length of the standing wave. Standing waves can be easily justified by a diagram like the one in Figure 8-11.

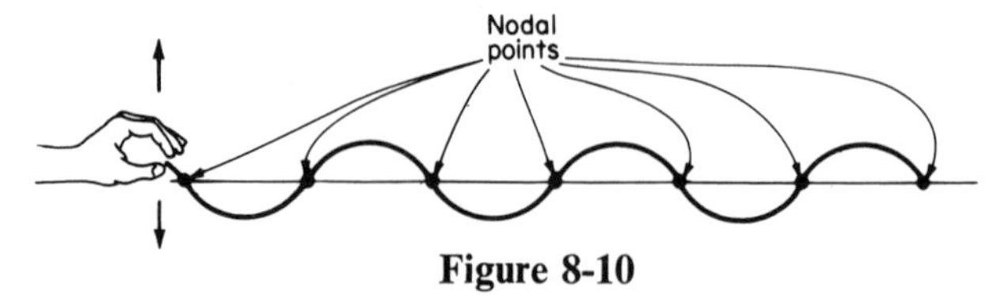

Figure 8-10

If we represent crests by the solid lines and troughs by the dotted lines, we can see represented that as a trough moves in to the fixed end it is reflected back as a crest. A node exists wherever a crest meets a reflected crest. In the diagram, Figure 8-11, $4\frac{1}{2}$ wave lengths are shown to exist between the two end nodes and ten nodes including the two end ones can be counted. One can also see that when the incoming wave and the reflected wave each move a quarter of a wave length, crest will meet crest and trough will meet trough and a maximum or anti-node will result.

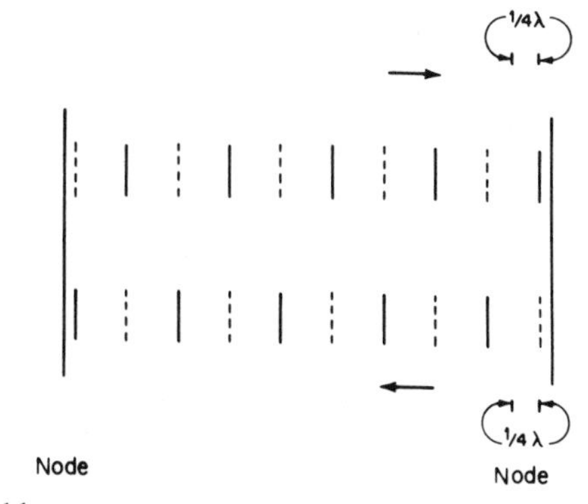

Figure 8-11

RESONANCE

At this point it would be useful to discuss the phenomenon of resonance. Introducing the subject here will help later in the discussion of the standing wave model for atomic electrons, resonance effects in atomic nuclei, etc. We have already discussed standing waves in musical strings. Using this discussion as a background, it is very easy to demonstrate resonance in musical strings.

Two musical strings are stretched side by side and tuned slightly differently. A small piece of paper is folded over one of them and the other one is plucked. The paper is seen to vibrate slightly. The string on which the paper is suspended is then tuned to match the other string, and the first string is plucked at intervals while the second string's tension is being changed. It becomes obvious that as the tuning becomes more and more matched, the second string vibrates more and more violently until finally when the first string is properly tuned, the second string vibrates so violently that the paper folded over it is shaken off.

Of course, the sound generated by the vibrating first string strikes the second string and disturbs it, but only when the pitch is right, when the frequency of the sound is such that it can accept it to produce a standing wave, will it accept the energy necessary to cause it to vibrate violently.

The analogy used to explain this might be that of a person pushing a child in a swing. If the pushes match the frequency of the swing, if you push down only when the swing is already going down, the amplitude of the swing is increased (its energy is increased). If, however, the frequency of the swing and the frequency of the pushes are mismatched, you might sometimes push down when the swing is coming up and the amplitude of the swing is lessened. (It does not accept the energy being given it.)

It would be useful to demonstrate this resonance effect in several ways to emphasize its universal nature. For instance, three tuning forks might be used, two of which are matched and one different. When two different ones are side by side and one is struck and then stopped, there is dead silence. When the two similar ones are side by side and one is struck and then stopped, a continuing hum is heard. The fact that this hum is caused by the induced vibration in the second tuning fork is easily demonstrated by stopping it and noting the silence.

We had previously done an experiment in which we found that the frequency of oscillation of a pendulum depends on its length. We can, therefore, "tune" a pendulum for a particular period of oscillation by changing its length.

A resonance demonstration using two coupled pendulums can be set up as follows: Two pendulums are suspended from a wire or string. One pendulum has a fixed length, the other, through an arrangement of slip knots, can have its length adjusted. Begin with the two pendulums of unequal length. It will be noted that when the pendulum with the fixed length is started swinging, the second pendulum vibrates slightly but does not accept the energy of the first pendulum. It just keeps swinging. However, when the second pendulum is properly tuned, when its length is arranged so that it matches that of the first, an unexpected (by the students) phenomenon is seen to occur. The first pendulum begins to lose amplitude very rapidly while the second pendulum begins to swing more and more violently with the same period. Finally, the first pendulum comes to rest and all the energy is in the second one; the second one begins to attenuate while the first starts swinging more and more violently. Energy is being passed back and forth from one pendulum to the other.

With this arrangement, one can also demonstrate harmonics. If the adjustable pendulum's length is arranged so that its frequency is a whole number multiple of the first (in the swing analogy, it would be like pushing down very second or third swing), the same effect is seen to occur.

The natural extension of these demonstrations is to point out where other resonance effects occur in nature. For instance, all are familiar with the process of tuning a radio or television set. Actually what they are doing is tuning an electric circuit so that it accepts only a particular desired frequency of the many available radio frequencies in the surrounding medium.

A qualitative discussion of absorption or Frauenhofer lines in the solar spectrum, wherein atomic electrons in the atmosphere surrounding the sun can accept only certain energies or frequencies radiated from the interior of the sun, will prove useful groundwork for fruitful discussion later.

The necessity for tying in the subject matter being currently studied with what has been previously discussed and with what is still to come must always be borne in mind if physics is to be taught as a science and not as a series of isolated facts.

9

Multi-Dimensional Waves

THE TRANSITION FROM STUDYING wave motion in mono-dimensional media to the investigation of wave phenomena in two dimensional media is smooth and natural.

The situation can be clearly seen when one disturbs the surface of a liquid and observes an expanding circular pulse emanating from the source of disturbance. The mono-dimensional situation may be seen to be a cross-section of the two-dimensional case in that if one looks along any selected diameter of the expanding circular pulse one sees two pulses moving away from a center of disturbance as in the mono-dimensional case. A three-dimensional pulse can then be likened to an expanding spherical bubble in which a cross-section would be the expanding circular pulse seen in the two-dimensional case.

The ripple tank and its accessories are marvelous instruments for studying two-dimensional waves.

Velocity of Waves

The easiest way to measure the velocity of waves in a ripple tank is to strobe the light with a hand stroboscope. The stroboscope is held between the light source and the tank and the light is strobed at the fastest rate necessary to stop the longest wave length image of the line waves being generated. The frequency of the waves is now known (by the frequency of the strobe wheel times the number of slits), and the wave length can be measured on the stationary image. The velocity is then wave length (λ) times the frequency (ν). The wave length measured is, of course, that of the projected image. To obtain the true wave length, and thus the actual velocity of the waves in the ripple tank, put a coin of known diameter in the tank and measure the diameter of its image. You thus obtain the ratio of sizes in the tank to their images.

Reflection of Two-Dimensional Waves

Experiments with single pulses involving circular pulses from a point source and linear pulses from a line source are very easy to do, and experiments involving reflection of single pulses from barriers can be performed by students completely independently. Perhaps one ought to point out that a straight wave represents a circular wave an infinite distance from the source. Most students can see that as a circle gets larger, its curvature becomes less.

To relate with what has already been done in geometrical optics, we ought to redefine a ray. In optics a ray was defined as the path along which light traveled. If we consider a particle model for light, the ray would be the path of a light particle. In the wave model, we must redefine a ray as the path of a point on an advancing wave front. The ray representing this path, therefore, is always drawn perpendicular to the wave front. Thus, if periodic waves emanating from a point source are represented as a series of concentric circles around the point, rays would be seen as radial lines originating from a common source.

One can predict then, by using the law of reflection learned in our optics experiment, how a wave should be reflected if the same law holds. (See Figure 9-1.)

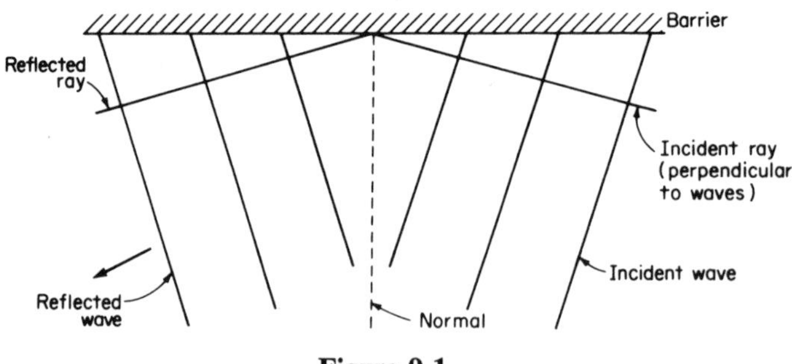

Figure 9-1

While the angle of incidence and the angle of reflection were previously defined as the angle between the appropriate ray and the normal, in the wave case it should be geometrically shown that these angles are represented by the angle between the wave front and the barrier or the interface if the wave is crossing over into another medium. (See Figure 9-2.)

Since ∢ a (the incident angle) is complementary to ∢ *b*, and ∢ *d* is also complementary to ∢ b, ∢ d must equal ∢ a, the incident ∢, since both are complementary to the same ∢.

Refraction of Two-Dimensional Waves

Refraction experiments are performed by placing a glass shoal in the ripple tank. Waves travel more slowly in very shallow water than they do in deep water. This can be experimentally verified by watching the change in wave length as the waves move across an interface from deep to shallow. One important thing to remember is that the difference in the speed of the wave,

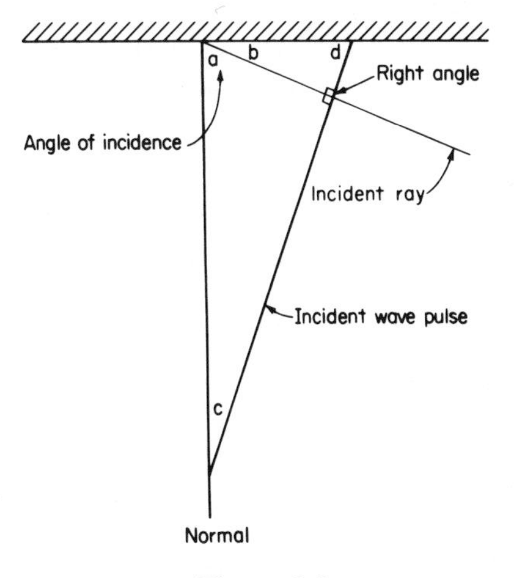

Figure 9-2

i.e., the difference in wave length, is only perceptible when the shallow area over the shoal is very shallow compared to the deep. The technique for insuring satisfactory results is to have your students raise the glass plate used for a shoal on coins or metal washers and add or remove enough water so the shoal is just barely covered.

Snell's Law can be easily derived by changing the angle at which the waves cross the interface (move the glass shoal around). Measure the angles of incidence and the angle of refraction (the angles made between the incident ray and the interface and the angle made between the refracted ray and the interface) and show that the ratio of the sines of these angles is always constant.

A useful technique for such observations and measurements is to have one of the students in a laboratory group spin a hand stroboscope between the light source and the tank so that the image thrown on the screen is strobed and held apparently motionless. Another member of the group can then trace the image of the incident and refracted wave fronts and the interface on a piece of paper placed on the screen being used and then leisurely measure the necessary angles.

One ought also to ask the students to measure the wave lengths in the incident and refracting media and show that the ratios of these wave lengths also equal the same constant, the index of refraction between the deep and the shallows.

Incidentally, as you have probably experienced, some students are easily confused by what seems to be the simplest of concepts. I was astonished one day to discover a group in one of my laboratories who thought they were measuring the index of refraction of light in water (because they were using water as a wave medium and light to project the image of these waves). They had to be carefully convinced that they were measuring the index of refraction of water waves going from deep water into shallow water.

An argument to show that the index of refraction is the same ratio for the sin $\sphericalangle i$ over sin $\sphericalangle r$, wave length in incident medium over wave length in refracting medium, and incident velocity over refracted velocity follows. (See Figure 9-3.)

$$n = \frac{\sin \sphericalangle i}{\sin \sphericalangle r} \text{ by Snell's Law}$$

$$n = \frac{\lambda i/X}{\lambda r/X} = \frac{\lambda i}{\lambda r}$$

$$n = \frac{\lambda i}{\lambda r} \times \frac{f}{f} \text{ (frequency is same in two media)} = \frac{Vi}{V_r}$$

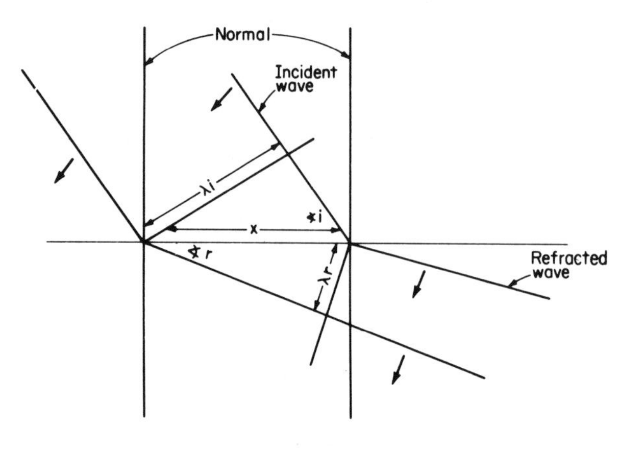

Figure 9-3

Interference

Interference of waves from two point sources offers a most satisfying group of laboratory experiments for students. Before starting these experiments in the laboratory, it would be well to discuss the interference effects they are about to see and to justify, by mathematical arguments, some of the relationships they will use in obtaining results.

Two point sources emanating circular waves can be represented by concentric semi-circles using solid lines for crests and dotted lines for troughs. (See Figure 9-4.)

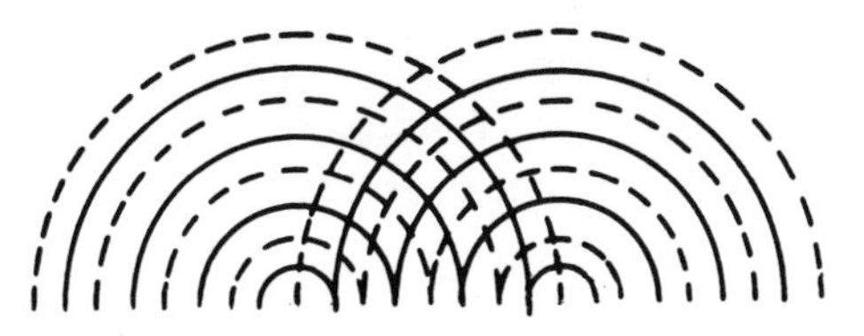

Figure 9-4

The nodal lines along which crest and trough constantly intersect can be clearly shown, and the lines of maxima where there is always constructive interference are also clearly evident and should help the student interpret what he will see in the laboratory. It would then be appropriate to isolate the sources and one wave for the diagram in Figure 9-5 and its interpretation.

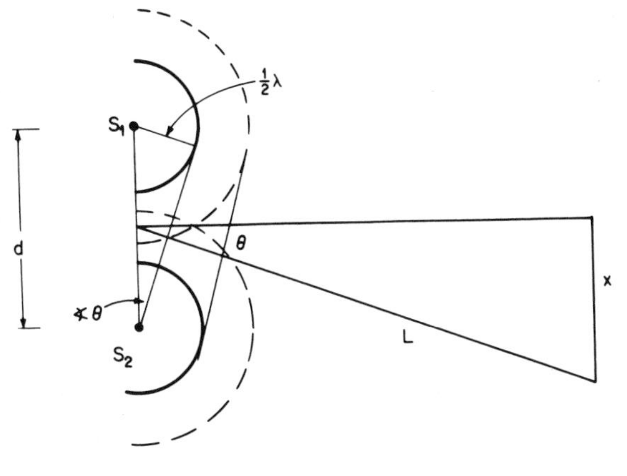

Figure 9-5

Any tangent drawn from the circular trough of one source to the circular crest either preceding it or behind it from the other source will mark a combined minimum wave front moving along the line L.

The two $\measuredangle\ \theta$'s shown in the diagram must be equal since they are both acute angles with mutually perpendicular sides. They are both in right triangles since $\frac{1}{2}\lambda$ is radial and must be perpendicular to the tangent drawn as the common tangent for the crest and trough and the line X is from a point on L to the perpendicular bisector of the line $S_i\ S_2$.

Therefore,

$$\sin \theta = \frac{\frac{1}{2}\lambda}{d} = \frac{X}{L}$$

Since the geometry would hold, if I drew the figure with a common tangent for the last trough having emerged from S_i and the crest just starting from S_2 (where $n = 1$), or if I similarly connected the trough preceding this last trough with the crest just emerging from S_2, then

$$\sin \theta = \frac{(n - \frac{1}{2})\lambda}{d} = \frac{X}{L}.$$

Where n is the number of the nodal line counted from the central maxima (the perpendicular bisector of $S_1 - S_2$), λ is the wave length, d is the distance between S_1 and S_2, X is the distance from a point on the selected node perpendicularly to the central maxima and L is the distance of that point on the nodal line from the midpoint of $S_1 - S_2$.

Once your students have digested this geometry, they can then proceed to measure the wave length of the water waves in an interference pattern using the expression

$$\lambda = \frac{d \cdot x}{(n - \frac{1}{2})L}$$

and then derive some satisfaction by shutting off one of the sources and confirming their result by direct measurement of the wave length.

One very important caution ought to be made here. The expression $(n - \frac{1}{2})\lambda/d = X/L$ is appropriate for measurements in the ripple tank because the nodal lines are so obvious. Later when this method is used for measuring the wave length of the various colors of light dispersed by a diffraction grating or a double slit as in "Young's Experiment," the expression used is $n\lambda/d = X/L$ because then X is the distance from maxima to maxima. This is not intuitively obvious to all students and must be pointed out.

The arguments are similar. Students should already have seen in diffraction experiments performed in the laboratory that a small slit (whose width is less than the wave length of the waves going through them) acts as a point source. (See Figure 9-6.)

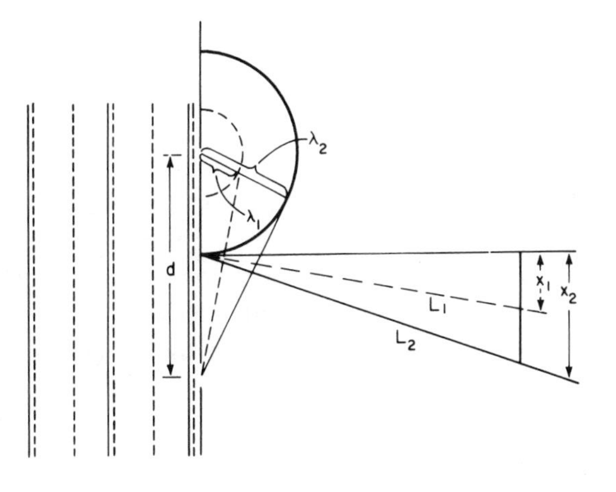

Figure 9-6

Straight waves of different wave lengths (white light) can be depicted as coming to the slit using solid lines and dotted lines or various colored chalks, and it will then become evident why the shorter wave lengths have maxima closer to the central maxima (smaller X) and the longer wave lengths fall further away from the central maxima.

Usually in spectroscopic analysis of light, you measure X to the first maximum so $\lambda/d = X/L$, since $n = 1$.

Sometimes some students are bothered by the fact that they see colored maxima on either side of the source when they know that dispersion is taking place on the side of the grating that their eye is on. It is necessary to explain that their eye follows the maxima reaching it to its apparent source, and only the real source would be on a line with the central maximum of all the colors coming to it through the grating. All the other colors would appear to emanate from sources on either side of the actual source. (See Figure 9-7.)

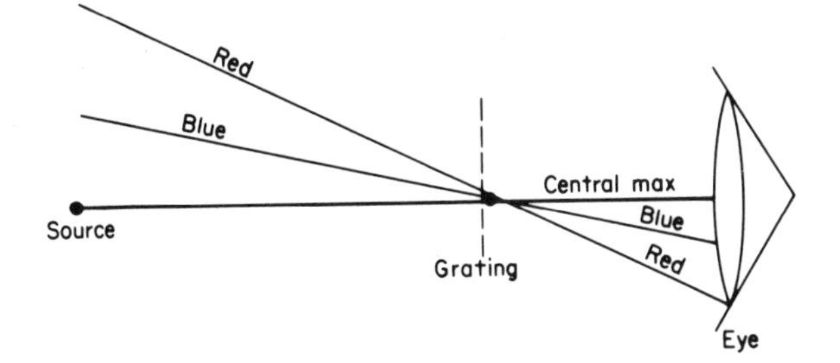

Figure 9-7

Phase

In some of the ripple tank experiments involving phase difference, a variable phase generator is used in which two point sources actuated by the same motor can be caused to beat the water surface at different times but at the same frequency. I suggest that the same effect can be demonstrated by using the regular two point source beating in phase. If one suspended the source so that one point is one-half wave length ahead of the other, the effect is exactly the same as if the two sources were both on the line of origin and one was producing a crest while the other produced a trough. In other words, one source one-half of a wave length behind the other, is like two sources in line with one 180° out of phase with the other, and a nodal line will now appear where previously the central maximum was.

Beats

An interesting demonstration which can then be related to other observable phenomena is the interference effect of two point sources that make disturbances of different frequency. Since the two sources will change phase with each other constantly, the interference pattern will be constantly shifting in a direction perpendicular to the propagation of the waves. The number of nodal lines going by a given point per unit time is then pointed to as the beat frequency.

Beat frequencies can then be demonstrated by two tuning forks of different pitch. The throbbing noise perceived is due to the nodes and antinodes alternately impinging on the ear. One might then point to a practical use of beat frequencies in the super heterodyning of radio frequencies for finer tuning.

An easy way to set up two oscillaters with different frequencies is illustrated in Figure 9-8.

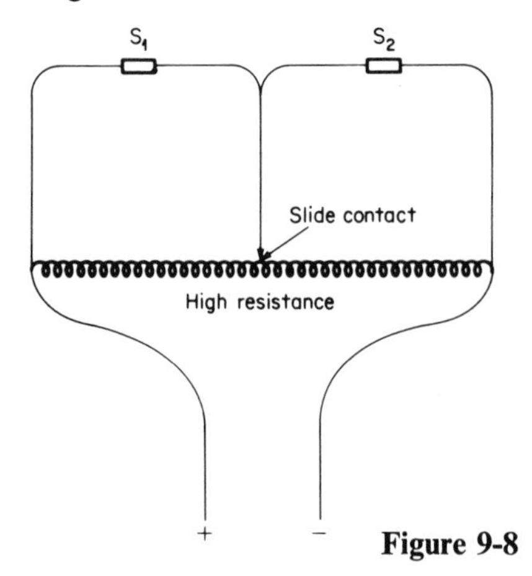

Figure 9-8

The slide contact is moved back and forth to raise or lower the individual frequencies of the two oscillators or to tune them to the same frequency, which becomes evident by the cessation of motion of the nodal lines in the interference pattern.

Another interesting demonstration, and one easy to accomplish with the ripple tank, is the demonstration of the "Doppler Effect."

I use a short piece of rubber laboratory hose through which I blow and alternately block and open the end in my mouth with my tongue to produce an oscillating point disturbance which I direct at the surface of the water with the other end of the hose.

The source can then be easily moved while the surface is disturbed and one can show a typical "Doppler Effect" pattern with the wave length shorter in the direction of motion and longer in the other direction.

The quantitative relationship for the apparent frequency observed by a stationary observer might be developed by the following argument: The speed of waves in the medium is constant (c). The speed of the disturbing source producing the waves might be assigned a value (v) as it approaches the observer and ($-v$) as it goes away.

Now let us assume that we are moving with the source and can look forward and observe the shorter wave length λ^1 and the waves moving away from us with a velocity $c - v$. We would then calculate the frequency ν to be $\nu = (c - v)/\lambda^1$. The observer who watches without moving and sees the waves approach him, would measure the frequency to be $\nu^1 = c/\lambda^1$ (he may not even be aware there is a moving source and he still observes the shorter wave length λ^1).

λ^1 then is equal to $(c - v)/\nu$ and c/ν^1.

These last two expressions can then be set equal to each other and we can solve for ν^1 in terms of ν.

$$\frac{c - v}{\nu} = \frac{c}{\nu^1}$$

$$(c - v)\,\nu^1 = c\nu$$

$$\nu = \nu^1\left(1 - \frac{v}{c}\right)$$

$$\text{and } \nu^1 = \frac{\nu}{1 - v/c}$$

If, on the other hand, you who are moving with the source, look backward and observe the long wave length λ^1 moving away from you with a velocity $c + v$, then similarly to above, the equation derived for the frequency when the waves appear to be longer as the source is going away from the stationary observer is $\nu^1 = \nu/1 + (v/c)$

All the above has been derived when the velocity of the source is less than the velocity of the waves generated. What happens when the source is moving faster than the waves it generates?

The situation is represented diagramatically in the sketch in Figure 9-9.

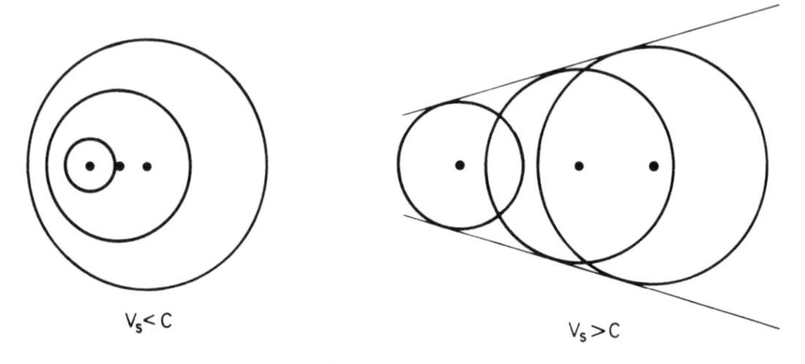

Figure 9-9

The wave fronts cross each to form a reinforced turbulence moving away from the line of motion of the source in the form a *V* shaped wake. All your students will have seen the wake moving from the prow of a boat when the boat is moving though the water faster than the waves it propagates. The sonic boom caused by fast flying military jets is an example of the same phenomenon. The planes are flying faster than the sound waves they generate and the "boom" is the *V* shape reinforced sound wave wake which is really a magnified rarefaction and condensation (analogous to the trough and crest of a transverse wave). This is the sudden compression and evacuation which produces the shock that breaks windows, etc. in the wake of a faster-than-sound airplane.

Another example which is not so familiar to students is the Cerenkov radiation seen in "swimming pool" reactors. Here the particles going through the water surrounding the nuclear pile are moving faster than the speed of light they are radiating in the water in which they are traveling.

This last brings up a question that a bright student sometimes asks which deserves some comment. We are told that the velocity of light in a vacum is a constant. Now any medium through which light moves is essentially empty space. A hydrogen atom, for instance, is approximately one angstrom in diameter, but its nucleus has a diameter of the order of $10^{-5} A°$. The physical dimensions of the electrons and protons making up the medium through which the light is moving is insignificant compared to the volume of the medium and, therefore, according to our questioning student, light is still traveling essentially in a vaccum. Why does it appear to slow down?

Some interesting discussion can ensue. A model favored by some physicists is called the extinction theory. It suggests that photons traversing an electron-containing media may be extinguished by passing their energy to an electron. The excited electron will quickly reemit a photon, but there is a time delay. Thus, the speed of light has not changed but has proceeded through the medium in a series of interrupted flights. This concurs with experimental evidence that the density of the medium affects this apparent "slowing down." Denser media would have more electrons available for extinction and reemission of photons and the photon would, therefore, have its motion interrupted more often.

Of course, on reemergence into a vacuum on the downstream side of the media, there are no longer any interruptions and the measured speed is again that of the speed of light in a vacuum.

Techniques

The source of light illuminating the ripple tank ought to be a point source. Usually a 150 watt line filament bulb is used. A point source is approximated by lining up the filament so that one end points toward the tank.

The reason for the desirability of point sources is that it is the images of this source which are projected on the screen by the crests and troughs of water waves which act like converging and diverging lenses. A line source arranged parallel to straight waves would also be suitable but would be unsatisfactory for circular waves, since a line, an infinite number of point sources, would cast many images and thus create a fuzzy pattern on the screen. The fact that water waves act as lenses would indicate that there are optimum distances between the light and the tank on one side and the screen and the tank on the other side for sharpest focusing. Students should be encouraged to use what techniques they can to produce most optimum experimental conditions.

Student set-ups are usually arranged with the light source above the tank and the screen below. A good demonstration set-up can be arranged by the teacher by putting the light source under the tank and projecting the image on the ceiling. If you have a white ceiling and can darken your room, the greatly magnified image of the wave pattern in the ripple tank can be seen by your entire class at once and you can point out all the things you wish them to see.

For interference effects of lights, I would suggest that a large roll of replica diffraction grating, 13,400 lines to the inch, can be obtained very inexpensively from Edmund Scientific Co. in Barrington, New Jersey. With this roll in your possession, you can cut off inch square pieces and give a square to each student, instructing him to frame it. 35 mm slide holders are very good for these frames. Each student can then own his own diffraction grating and be encouraged to do spectroscopic measurements at home.

I have also cut off a larger piece of this grating, about 6″ × 6″, and framed it with a wooden frame. With this piece of equipment I can point out spectral lines from a line source to a large group of students at once and they can all see what I want them to see. This facilitates such demonstrations.

An effective lab test after completing diffraction and interference experiments in ripple tanks and doing the standard Young's experiment with homemade double slits is to set up several line sources around the lab (I usually set up six). These are usually fairly simple spectra sources like Helium, Hydrogen, Sodium, and Mercury. I then ask my students to use the gratings I gave them and a pair of meter sticks to measure, with what accuracy they can, the wave length of at least two different spectral lines at the station I assign them. They use one sheet of standard composition paper, on one side of which they report their measurements and calculations which are graded for precision in measurement and accuracy of results. On the other side, they write a short essay explaining theoretically how their diffraction gratings made it possible to make such measurements.

Wave Model vs. Particle Model

In building a particle model for light one can easily show that particle reflection obeys the laws of reflection. Refraction effects can be shown by starting a ball rolling on a plane at some angle to an interface with lower surface. The trouble is, of course, that it bends the wrong way. Newton justified this bending of particles by the following logic. (See Figure 9-10.)

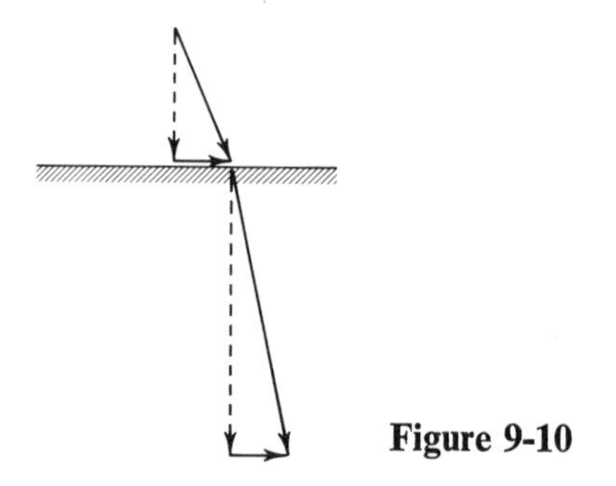

Figure 9-10

When a particle with a given velocity crosses an interface, the increase in velocity due to the crossing is given only to the normal component of its velocity. (There can be no crossing except in the normal path; the parallel component to the interface never crosses it.) Therefore, as illustrated in the diagram above, the particle bends toward the normal when it crosses to a faster medium, and the index of refraction would be the reciprocal of what is found in experiments with light.

The demonstrations with waves in ripple tanks, showing proper refraction, interference effects, and diffraction make strong arguments for the fact that light appears to behave as if it were a wave. In fact, demonstrations with polaroid make it evident that light can be polarized and must, therefore, behave like transverse waves. With all this overpowering evidence, one must still say that light appears to behave like a wave. We still cannot, and must be careful not to, infer that light is a wave. Later we will find evidence that in some aspects of its behavior only a particle-likeness would explain some of our experimental results, and we are left with the conclusion that we really don't know what light is.

At this point, a teacher can point out to his students that man is subject to a very small spectrum of the possible universal band of experiences and he always tries to interpret everything he sees, hears, and feels in the light of his own experiences. A simple analogy might be drawn of the Australian aborigine who sees an airplane for the first time. He watches it land, goes over and touches it, and then watches it take off and fly away. He then goes back to his colleagues at his tribal home and tries to explain what he saw.

"It flies like a bird, is cold and inanimate like a stone." He cannot communicate. There is nothing in the experience of his cohorts which permits them to relate.

10

Newton's Laws of Motion

NEWTON'S LAWS OF MOTION ARE THE BASIS for the study of classical physics. Conventionally this is usually taught as (1) the law of inertia, "Every material body persists in its state of rest or of uniform motion in a straight line unless acted upon by an external, unbalanced force," (2) $F = MA$, the change in motion or the acceleration is directly proportional to the applied net force and indirectly proportional to the mass or inertia, and (3) every action has an equal and opposite reaction.

Actually, there is only one law and that is the law of inertia. An object with a given mass resists a change in motion, and the amount of such resistance to change is proportional to the amount of mass. The second law is really a corollary to the first. If it takes a force to cause this change in motion, the amount of change is proportional to the amount of applied net force. The third law evolves naturally from the second law.

If $F = ma$, then $F = m\Delta V/\Delta T$ and $F\Delta T = m\Delta V$. Force exerted for a given length of time (the impulse) gives a mass a specific change in velocity (momentum change ΔR). If the law of inertia holds, this implies that momentum is conserved, and conservation of momentum predicts that every action is accompanied by an equal and opposite reaction. If an isolated system, having no momentum with respect to an observer and having no external net force applied to it, is caused to move due to some internal force, then any momentum in one direction must be accompanied by an equal and opposite momentum in order for the total momentum to remain zero.

Having the student discover this for himself by leading him along Galileo's path to the same conclusions regarding inertia would certainly be more meaningful than just having him memorize the above quotation for Newton's first law.

It has already been ascertained (in Chapter 4 on Kinematics) that a body starting from rest and falling toward the earth will accelerate at a constant rate. It can also be demonstrated that objects rolling down inclines will accelerate at rates depending on the slope of the incline.

Similarly, students can be led to see (by knowledge born of their own lifetime experiences) that a ball rolling up hill will gradually decelerate. One can argue, then, that a body neither rolling up nor down, that is, moving on a level surface, must neither increase nor decrease its velocity and will continue to move at constant velocity forever unless some force acts on it.

The fact that in actual practice it does gradually slow down and come to a halt can be used to develop the idea of friction as a retarding force.

Experiments and demonstrations showing that balls rolling down a ramp (see Figure 10-1) will roll up almost as high on the other side regardless of the slope, can be very useful. As the slope of the up side is made less and less, the ball is seen to travel further and further in order to reach its original height.

This is reputedly Galileo's method of demonstrating the law of inertia. The PSSC laboratory guide suggests an adapation of this experiment. A pendulum moving through a half cycle is essentially falling down a hill and climbing up again. By using a very long pendulum and placing a barrier in the way of the string, the Galilean experiment of steep slopes and gentle slopes is accomplished. If you use this experiment, make sure all your students see this. It has been my experience that many students fail to make the connection without some help. (See Figure 10-1.)

In the laboratory development of $F = ma$, one might first keep the mass constant and pull the mass with varying forces measuring the resulting acceleration. Acceleration can be measured with the timer and tape described earlier in the experimental derivation of $D = \frac{1}{2} at^2$. It would then be found, by graphing the force vs. the acceleration, that a linear relationship has been attained and force is proportional to acceleration if the mass is held constant.

To minimize the effect of forces other than the applied force, make sure the surfaces on which your students work are as level as you can make them and that friction is minimized by using well-lubricated ball bearings or any other means of cutting down on friction.

The forces applied may be as suggested in the PSSC lab, rubber bands

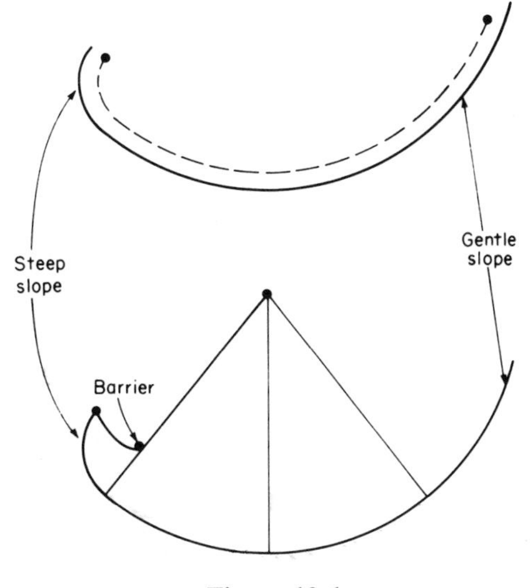

Figure 10-1

stretched a given amount and then using one, two, three, four, or five of these all stretched the same amount. Some teachers prefer pulling the masses with spring scales using the stretch of a Hooke's Law spring to measure the forces.

It is even possible to use a weight over a pulley to pull the mass, but here a strong caution must be interposed. The weight must be very small compared to the mass being pulled in order to have it contribute a negligible amount of inertia.

If this method is used, and it has its advantages since the force will be much more constant if gravity is doing the pulling than when the student is applying the force, it should still be preceded by a few runs in which the student applies the horizontal force. This is necessary in order to emphasize the concept that the force being measured is force in general and not gravity, and that if gravity is used as the experimental cause of the force, it is just one other way of applying a force.

To emphasize this point, it might be well to place a spring scale in the horizontal portion of the line between the weight and the mass and have the student read his applied forces from the scale.

The next obvious step or second experiment is to keep the force constant and vary the masses being pulled. Upon graphing the mass vs. the obtained acceleration with constant force, a hyperbola indicating an inverse relationship is attained. This suggests graphing the mass vs. the reciprocal of the acceleration and a linear relationship indicating that $m = k/a$ is seen.

We now have experimentally discovered that $k_1 = ma$ and $F = k_2a$.

These can be combined to read $F = ma$ which says that when $F = k$, $k = ma$, or when m is held constant $F = ka$.

If the units of mass are measured in kilograms and units of acceleration are in meters per second per second (m/sec^2), then the unit of force must be the product of these ($Kg \cdot m/sec^2$) which is called a newton.

Two very important concepts should have made themselves clear in the experiments outlined above. One is that acceleration must be caused by a force and must be proportional to that applied net force. The other is that mass is a measure of the amount of inertia a body possesses. In fact we may use it as another definition for mass. Previously we have defined mass as a quantity of matter. Now we can also define mass as a quantity of inertia and measure both matter and inertia in units of kilograms.

Action and reaction is most meaningful, I believe, if it is taught as conservation of momentum. A rifle at rest has no momentum. When the powder in the cartridge is exploded, no external force is added and the total momentum is still zero. The momentum given the bullet and the escaping gases must, therefore, be matched by a momentum equal in quantity but opposite in direction (the kick of the rifle).

Experiments involving two ball-bearing carts with a tightly wound spring between them which is permitted to explode when triggered are very good for emphasizing this concept. The student measures the resulting momentum by measuring the distances moved by the two carts in equal times and multiplying the distances by the masses of the propelled carts.

The time factor is held constant for the two carts by placing barriers on the table so that by listening for the sound impact between the carts and the barriers, the student can place the barriers appropriately so that the impacts are simultaneous. Again make sure that the table surfaces are absolutely level and the wheel bearings are well lubricated.

Actually, in doing this experiment, the student is locating the center of mass before and after the explosion and, if momentum is conserved, he will find that the center of mass between the two carts has remained at rest. It would be well to emphasize this center of mass concept. It is a useful way of looking at such problems and will help your students visualize what is happening when they are required to solve more complicated problems involving momentum conservation.

It goes back to what we call Newton's first law. If the center of mass of a system is at rest, it remains at rest, or if it had some velocity before the internal explosion it continues to have the same velocity until some external net force is applied to the system.

This can be further simplified. You might point out that when you say an object is at rest you are implying that it has a constant velocity. An object at rest is only at rest with respect to an observer who is traveling at the same velocity as the object. It is certainly not at rest to the observer traveling at some different velocity.

The standard example for the third law is rocket propulsion. Gases escaping from the tail of the rocket must cause the head of the rocket to move forward. One way to have your students visualize this concept is to suggest that the rocket might be full of holes in all its walls so that the gas inside can then expand in all directions without moving the rocket. Suppose the holes were only in the tail. The expanding gas can then escape in that direction, but the solid forward wall would experience an unbalanced net force and the rocket would be accelerated in that direction. This is a satisfactory qualitative approach to the concept, but, it seems to me, dealing with rocket propulsion from the conservation of momentum viewpoint permits a more quantitative approach to the subject.

A useful demonstration of inertia may be arranged by hanging a heavy mass from a very light cord and suspending a similar light cord below the mass.

If you pull on the suspended cord with a hard sudden jerk, you can apply enough force for a very short time to overcome the cohesive forces holding the molecules of the cord together and the cord will break below the suspended weight. The cord above the suspended weight is protected by the inertia of the suspended weight. When, however, you pull on the lower cord with a gentle force (not enough to overcome the tensile strength of the string) the string will eventually break above the weight. The small force applied for a long time is sufficient to supply a momentum to the mass hanging on the string ($F\Delta T = m\Delta V$). Once sufficient momentum is accumulated by the mass, the inertia of the mass causes the string to break above the mass since the cohesive forces are so small that there is not enough time to retard the mass before the string is stretched beyond the breaking point.

If you wish then to make this point a bit more dramatic, you can hold a heavy mass on your head (an anvil or a cinder block is satisfactory). A student can then be invited to hammer a nail into a block of wood on top of the mass resting on your head. The large force imparted for a short time to the nail does not impart sufficient momentum to the inertial mass resting on your head to cause you any discomfort.

A most dramatic variation of these demonstrations is one I have seen performed by some of the street magicians in India. The magician lies on his back while he holds a massive stone (about 75 to 100 lbs.) on his chest. He then challenges onlookers to pick up a sledge hammer and try to break the stone while he holds it on his body. Again it is the inertia of the stone that prevents the impulse delivered by the hammer from being transferred to his chest. He also keeps the stone slightly suspended above his chest by supporting it on his hands so that immediately following the impact he can partially retard its downward progress with his arm muscles, thus minimizing its actual contact with his body.

We have thus far discussed momentum conservation in an explosion. It must be equally true that momentum is conserved in collisions. (After all, a collision is just an explosion run backwards in time.) It would be wise to defer the discussion of collisions until after a discussion of energy and its conservation.

Instruct two students to stand on skate boards. Select two students of contrasting mass. When they push or pull on each other, it will be observed that the lighter student moves much more than the heavier student. It is possible to make some quantitative calculations showing that the center of mass between them remains stationary. One might follow this by having one of the students push against a fixed object like a wall in the classroom and ask someone to explain his apparent center of mass displacement.

The answer, of course, is that the wall is fixed to the building and the building is fixed to the earth, and the displacement of the earth due to his push must be infinitesimal due to the relative infinite mass of the earth. The center of mass between the student and the earth must remain fixed even though the motion of the earth is not obvious. This must happen every time we walk. As we propel ourselves forward, we must kick the earth backwards just a little. Usually when this is proposed, someone suggests that there is someone in India or China walking toward us so that he helps keep the spin of the earth stable.

The effect of walking can be demonstrated by having a student walk on a platform with ball-bearing wheels freely turning so that it is not coupled by friction to the earth. It will be observed that as the student walks forward, the platform beneath him is pushed backward.

Another variation of this might be to mount a pendulum on a wheeled cart. As the pendulum swings, the cart is seen to move back and forth in the opposite direction, making it obvious that the common center of gravity is obeying the laws of inertia.

Action and reaction is also demonstrated by the classical Hero's engine. I have made these from a tin can with holes bored in the side with tangential spouts soldered over the holes. The apparatus is then filled with water (by immersing one of the spouts in water and sucking in on a spout on the opposite side) and hung from a fishing line swivel (a hook has been soldered to the top of the can). A bunsen burner is used as the heat source for boiling the water in the can. The front row of students usually gets sprayed a little in this demonstration.

Two pendulums mounted side by side, lifted to the same height in opposite directions and allowed to fall and strike each other, will also demonstrate the stability of their combined center of mass while they themselves are in motion. This can be further emphasized by coating the striking faces with plasticene or soft sponge so that the collision becomes inelastic.

11

Periodic Motion

IN THE INTRODUCTION of the physics of gravitational systems, some reference to historical background would be appropriate. For outside reading, books like *The Birth of a New Physics* by I. Bernard Cohen, *The Watershed, A Biography of Johannes Kepler* by Arthur Koestler, and *Sir Isaac Newton, His Life and Work* by E. N. da C. Andrade should be strongly recommended and, I am sure, would be enjoyed by most of your students.

The evaluations of man's ideas about his cosmic environments from Aristotle and Ptolemy, through Copernicus, Galileo, and Kepler, and finally all tied together by Isaac Newton makes fascinating reading. A teacher can also use illustrations here as a sounding board to emphasize the difference between objectivity in science and conclusions based on bias and misconception.

Galileo's "Abjuration" should be cited. The warning that a similar role can be played by fanatics and willful men today can be illustrated by the pseudo-scientific racist doctrines of Nazi-Germany, Lysenko's success in reviving Lamarkian biology in Russia because it fitted the "party line," and the racist literature coming from segregationists. All indicate the need for recognizing bias in oneself and in the arguments advanced by others.

While historically Newton's conclusion in the law of universal gravitation is based on Kepler's Laws and Kepler was influenced by Tycho Brahe's observation and the writings of Copernicus, it might be interesting to your students to show that they can use what they now know to predict what Kepler had arrived at by painstakingly fitting together minutia.

For instance, they have learned that $F = ma$ and that an expression for centripetal acceleration is $a = 4\pi^2r/T^2$. (Such an expression could also have been justified by the suggested lab exercise on circular motion.)

Newton's Law of universal gravitation is $F_g = GMm_s/R^2$.

Now when a satellite revolves about a very large mass, the centripetal force that keeps it moving around must be the gravitational force and

$$F_g = F_c$$

$$\frac{GMm_s}{R^2} = \frac{4\pi^2Rm_s}{T^2}$$

The mass of the satellite, therefore, is cancelled out since it occurs on both sides of the equation and

$$\frac{GM}{R^2} = \frac{4\pi^2r}{T^2},$$

thus equating the gravitational acceleration to the centripetal acceleration which is the same irrespective of the mass of the satellite.

By dividing both sides by $4\pi^2$ and multiplying both sides of the equation by R^2, one arrives at $R^3/T^2 = GM/4\pi^2$. This, then, is Kepler's third law which he derived by laborious comparison of the data accumulated by himself and Tycho Brahe on the orbital motions of the planets.

$GM/4\pi^2$ is the constant in his third law since G is a universal constant, M is constant for any satellite revolving around that particular M and, of course, $4\pi^2$ is just a number.

Now $GM_{sun}/4\pi^2$ is certainly not the same constant as $GM_{earth}/4\pi^2$. This would indicate that R^3/T^2 for a satellite of the earth is not the same as the R^3/T^2 for a satellite of the sun.

It is this argument which finally clinches the old problem regarding the revolving of the sun around the earth or the earth around the sun.

This constant R^3/T^2 for Earth, Mars, and Venus around the sun is the same as it is for any other satellite of the sun. The R^3/T^2 for the moon or any of the numerous artificial satellites now in motion around the earth is a different constant. In fact the equality of the gravitational force to the centripetal force can be used to determine the mass of the body being circled.

For instance, suppose we hear a news broadcast in which we are told that a new satellite has been placed in orbit. We are told that it makes a complete revolution every 98 minutes and that its maximum distance from the earth is 412 miles and its closest approach is 210 miles. Its average distance from the surface of the earth, therefore, is 311 miles, and when we add the radius of the earth to this (311 + 3,969) we get a value of 4,280 miles or 6900 km as the average radius of the satellite's orbit around the earth.

Since we know that

$$\frac{GM_eM_s}{R^2} = \frac{4\pi^2RM_s}{T^2},$$

substituting the values for the known parameters, we obtain

$$M_e = \frac{4 \times (3.14)^2 \times (6.9 \times 10^3 \times 10^3\ M)^3}{6.670 \times 10^{-11}\ \frac{M^3}{\text{Kg} - \text{sec}^2} \times (98\ \text{min.} \times 60\ \frac{\text{sec.}}{\text{min.}})^2}$$

and the mass of the earth can be calculated to be about 5.5×10^{24} Kg. Since the actual mass of the earth as given in astronomical tables turns out to be closer to 5.9×10^{24} Kg, we find we have come within 10 per cent of the agreed figure. For a student, the ability to calculate the mass of the earth within 10 per cent accuracy from the data given him by a newscaster, should be a source of pride to him.

Periodic motion, and especially that periodic motion called simple harmonic motion, can be analyzed now and demonstrated with little difficulty.

An object moving in a circle at constant speed is undergoing a periodic motion. The projection of its motion back and forth along a diameter of its circular path is the form of periodic motion called simple harmonic motion.

The diagrams that follow will serve to illustrate how such motion can be analyzed.

By indicating the components of the velocities in the path which are perpendicular and parallel to the diameters, it becomes easy to illustrate that in simple harmonic motion the velocity becomes zero at the two extremes of the motion and is greatest in the center. (See Figure 11-1.)

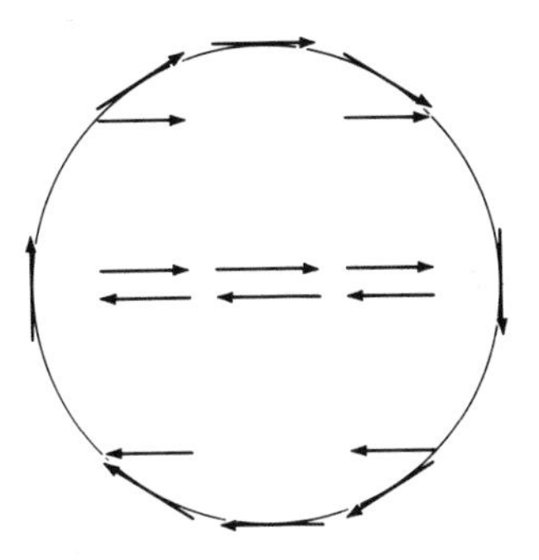

Figure 11-1

The acceleration in circular motion points radially back to the center of the circular path. By indicating the components of the circular acceleration perpendicular to and parallel to the diameter of the circular path, it becomes obvious that acceleration in simple harmonic motion is greatest at the extremes of the motion, zero in the center, and always points to the center of the motion. (See Figure 11-2.)

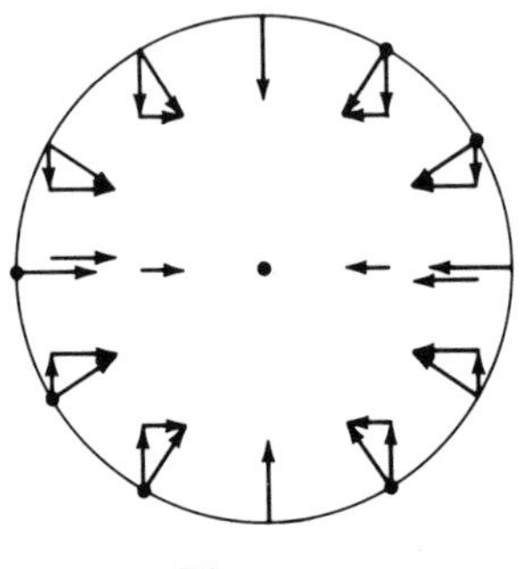

Figure 11-2

A turn-table is placed on the demonstration table with a brightly colored knob on an edge. The turn-table is then rotated at a constant speed so that the students seated in front of the demonstration view the circular motion of the colored knob as a projection of its motion back and forth along the diameter of the circle.

A pendulum can then be suspended above the turn-table and its length adjusted so that its period matches that of the turn-table. The frequency of the pendulum can then be arranged to be in phase with the motion of the turn-table knob as viewed by the students, and the exactly matched motions of the knob and the pendulum can be observed. (See Figure 11-3.)

A demonstration illustrating that in a pendulum moving back and forth in simple harmonic motion, the acceleration is greatest at the ends of the

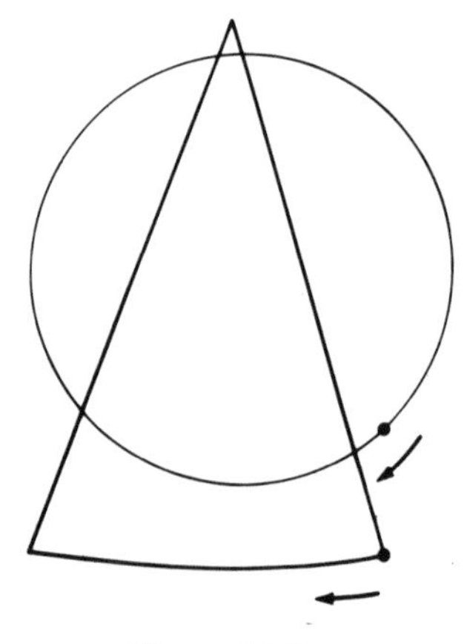

Figure 11-3

motion, zero in the center and points back to the center of motion can be arranged using the simple accelerometer described earlier. (See Figure 11-4.)

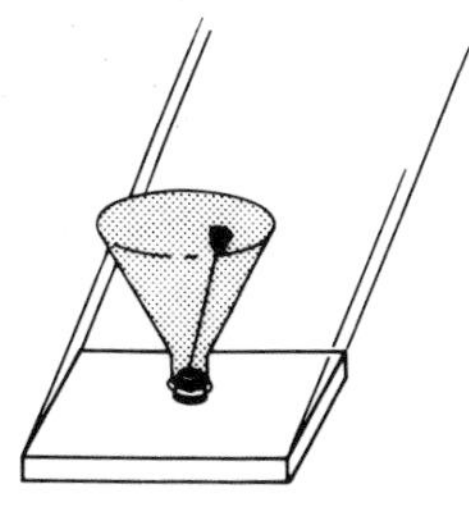

Figure 11-4

A platform swing type of pendulum is arranged and supported by parallel strings so that as the pendulum moves back and forth the platform remains level. The accelerometer is placed on the platform and the platform is set in motion. The accelerometer is seen to point always to the center of motion, pointing directly upwards when it is at the center of motion (indicating zero acceleration), and pointing toward the center with the greatest angle of deviation from straight up at the two extremes of its motion.

The fact that oscillating springs also obey the laws of simple harmonic motion can be similarly demonstrated. With the same turn-table mounted in front of the class, a ball-bearing wheel cart can be set in front of the turn-table. The cart is attached to a long soft spring at each end and the ends of the two springs are attached to opposite ends of the demonstration table. (See Figure 11-5.)

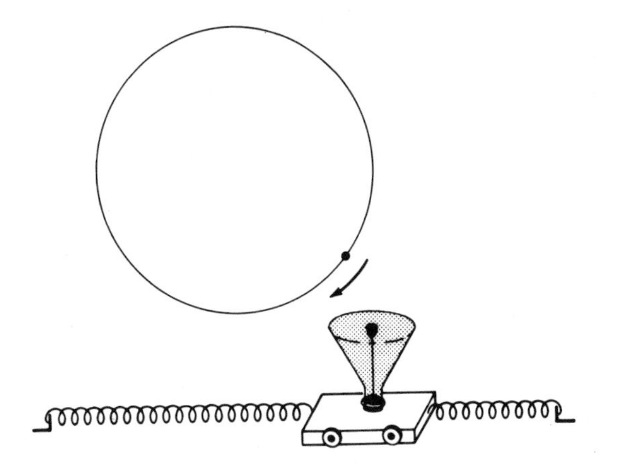

Figure 11-5

The tension of the springs can be adjusted so that the period of the oscillating motion of the cart is adjusted to that of the turn-table. The motions are then started in phase and the observation is made. (The motion of the pendulum suspended above the turn-table can be observed simultaneously.)

The accelerometer placed on the cart and undergoing the repetitive motion of the cart will again illustrate the role of acceleration in simple harmonic motion.

It might be useful here to point out that this could have been predicted since a spring obeys Hooke's Law where $F = kx$ (this might have been previously demonstrated to be true), and one sees that the force is proportional to the magnitude of the stretch of the spring. Being a restoring force, it points opposite to the direction of the distortion. Since the acceleration must be exactly proportional to the force (if the masses remained constant), then the prediction made by Hooke's Law is in exact agreement with what you previously had shown in your S.H.M. demonstration and you might now redefine simple harmonic motion as that form of periodic motion in which the force is a Hooke's Law type of recovery force.

The information accumulated through these presentations and demonstrations can then be summarized and extrapolated as follows:

It has been shown that the centripetal force causing an object to move in a circular path is $F_g = 4\pi^2 rm/T^2$, or solving for the period T, $T = 2\pi\sqrt{rm/F}$.

The period of an oscillating spring would be exactly the same except that we will use x (the extension of the spring) as the component of r along the selected diameter parallel to the spring motion. Then we can say that $T = 2\pi\sqrt{xm/F}$. But we have shown that the force activating the spring is a Hooke's Law force and $F = kx$. We can substitute kx for the magnitude of F in the equation for the period and $T = 2\pi\sqrt{xm/kx} = 2\pi\sqrt{m/k}$.

The magnitude of the force is pertinent; its direction here is not since the period T is a scalar quantity.

Confirming experiments can then be scheduled for students in which they can measure the spring constant by finding the slope of a F vs. x graph determined by hanging various weights on a spring and measuring the resulting distortion. The period can then be predicted for various masses and confirmed by experiment. One must caution that the mass of the spring must be negligible as compared to the mass being oscillated by the spring since the distortion (x) must be measured to the combined center of mass.

The relationship thus derived and confirmed can be used to derive the equation for the period of a pendulum.

In a pendulum it can be seen that the only force causing the motion is the weight of the oscillating object (again we make sure that the string supporting the weight is relatively massless with respect to the weight). (See Figure 11-6.)

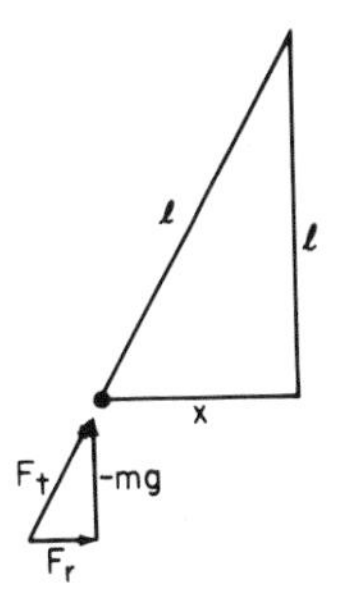

Figure 11-6

$$\frac{F_r \ (recovery\ force)}{M_g \ (weight)} = \frac{x}{l}$$

For S.H.M., $F_r = kx$

then $\frac{kx}{mg} = \frac{x}{l}$ and $k = \frac{mg}{l}$

substituting in the equation derived for S.H.M. in springs

$$T = 2\sqrt{\frac{m}{k}} = 2\sqrt{\frac{m}{mg/l}}$$

Then the period of a pendulum undergoing S.H.M. can be expressed as $T = 2\sqrt{l/g}$

Please note that this is only true when the angle of oscillation is very small since l must be constant through the motion of the pendulum mass along x and this is only approached for very short distances along x (the horizontal component of the pendulum's motion).

An interesting home assignment for students now would be to have them determine the gravitational field near the surface of the earth using a pendulum and the relationship just derived for the period of a pendulum.

The dependency of the period of the pendulum on gravity can also be demonstrated as follows:

Make a pendulum with a fairly heavy piece of ice or dry ice and suspend it in the plane of a large piece of glass (one of your ripple tanks would do nicely). The glass plane can then be tilted at various angles with respect to earth, and the resulting change in the oscillatory period of the ice pendulum can be observed and measured. Quantitative measurements relating the period to the angle of tilt can be made to confirm the effect of a lighter gravitational field. One can use such a demonstration to illustrate how a pendulum might behave near the surface of the moon. (See Figure 11-7.)

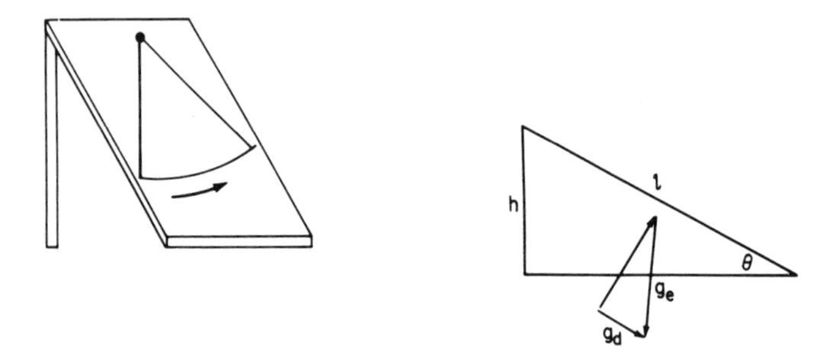

Figure 11-7

Since g_e is always downward, a tilted table counteracts part of g_e (the gravitational field of the earth). Then g (d), the diluted fraction of the field still acting on the pendulum, can be found by the equation

$g(d)/g_e = h/1$ or $g(d) = g_e \sin\theta$ where θ is the angle of tilt.

Then the period of the pendulum in a tilted plane is $T = 2\pi 1/g \sin\theta$ and as θ approaches 90°, $\sin\theta$ approaches one.

Gravitational Mass vs. Inertial Mass

An interesting discussion might ensue if you ask your students if the mass exemplified in the formula GMm/R^2 is the same as that in Newton's Second Law, $F = ma$. It is usually accepted without question, yet in the one case we are measuring the amount of attractive force which exists between two bodies and in the other we are concerned with the amount of resistance to a change in motion a body has. It turns out experimentally that the gravitational mass and inertial mass of any body is always proportional, but the fact that this is so does not necessarily explain why it is so. A series of experiments by Eötvos, and more recently by Dicke, establishes the proportionality of these two properties to many significant figures. By using the proportionality constant G in the gravitational equation, we can refer to the number of kilograms of gravitational mass and substitute for the same number of kilograms of inertial mass in $F = ma$ and not even wonder why this can be so without having it pointed out.

It is good for students to be questioned about and be made to realize the existence of another case wherein we accept the experimental evidence and use knowledge so gained with such confidence that very few of us have ever stopped to wonder why we can do so.

It is the equivalency between inertial and gravitational mass that forms the basis for Einstein's Theory of General Relativity.

Angular Momentum

Kepler's Law of equal areas should be stressed as an illustration of the conservation of angular momentum. The area of the triangle swept out by the displacement ($\vec{R}$) from the force source to the path is $\frac{1}{2}$ the altitude multiplied by the base. (See Figure 11-8.)

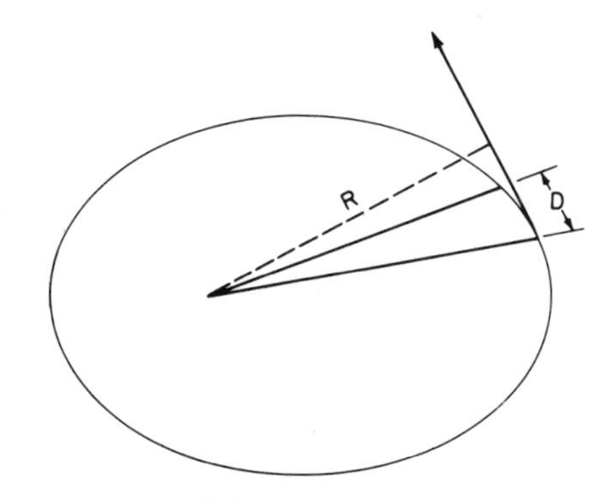

Figure 11-8

The base of the triangle is the distance traveled in the orbit during the time Δt. The altitude of the triangle is the component of the radius or displacement vector perpendicular to this base. Since according to Kepler's second law, $A/\Delta t$ is a constant, it therefore follows that $\frac{1}{2}DR_{\perp}/t$ or $R_{\perp}\ D/T$ is constant. D/T is the velocity (v). If $R_{\perp}\ V$ is constant and we multiply this quantity by the mass (m) of the rotating object, we have shown that the quantity $m\ R_{\perp}\ v$ is conserved, and we define this quantity as the angular momentum.

12

Work, Energy and Power

IT IS DIFFICULT to define work without using the word energy, and yet how do we define energy without saying that it measures the ability to do work?

The concept might be introduced by pointing out a kinematic relationship we had previously derived:

$v^2 = 2ad$

If we multiply both sides of this equation by the mass of the body being observed, we get

$mv^2 = 2mad$ or

$\frac{1}{2} mv^2 = mad.$

Since $F = ma$ (as in Newton's second law), we get $\frac{1}{2}mv^2 = F \cdot d$ and we call one side of this equation ($\frac{1}{2} mv^2$) the amount of kinetic energy the body possesses and the other side ($F \cdot d$) the amount of work necessary to have been performed on it to give it this kinetic energy, and then we do a series of experiments to verify this prediction.

A useful demonstration can be set up by mounting a pair of strong magnets on the carts used in the $F = ma$ experiments. The large magnets which can be removed from wave guides obtained from naval surplus depots are very good for this purpose.

Potential energy can be demonstrated in a collision as one sees a cart moving toward the other with evident kinetic energy, reach a point when they are closest to each other and seemingly lose a great deal of this kinetic energy, and then apparently regain this energy as they get further away from each other again.

Emphasize this point by showing how hard one has to work in order to push the two repelling magnets close to each other and then by releasing the magnets indicate how potential energy becomes kinetic energy.

Vary the mass of the incoming cart and demonstrate that when the incoming cart is less massive than the target cart, the incoming cart will recoil. When the incoming mass is equal to the target mass, the incoming mass will stop and the target mass will go off with same velocity the incoming mass had. When the incoming cart is more massive than the target mass, the incoming mass will continue with a small velocity after the collision while the target mass will move off with a comparatively high velocity.

All this can then be justified by the laws of conservation of energy and momentum.

If before the collision, all the energy and momentum is in the incoming mass, then

$$\left.\begin{array}{l} m_1v_1 = m_1v_1' + m_2v_2' \\ \text{and} \\ \frac{1}{2}m_1v_1^2 = \frac{1}{2}m_1v_1'^2 + \frac{1}{2}m_2v_2'^2 \end{array}\right\} \text{conservation laws}$$

dividing through by $\frac{1}{2}$, subtracting $m_1v_1'^2$ from both sides, we get

$$m_1v_1^2 - m_1v_1'^2 = m_2v_2'^2$$

factoring,

$$(m_1v_1 - m_1v_1')\,(v_1 + v_1') = m_2v_2'\cdot v_2'$$

but $m_1v_1 - m_1v_1' = m_2v_2'$ (from conservation of momentum) therefore $v_2' = v_1 + v_1'$

substituting in the expression for conservation of momentum we get

$$m_1v_1 - m_1v_1' = m_2v_1 + m_2v_1'$$

and

$$m_1v_1 - m_2v_1 = m_2v_1' + m_1v_1'$$

therefore

$$v_1' = \frac{(m_1 - m_2)}{m_1 + m_2}\,v_1$$

Inspection of this equation shows that if $m_1 = m_2$, v_1' (the velocity of m_1 after the collision) becomes zero. If m_1 is greater m_2, then v_1' will be smaller than v_1 but will still be positive, that is, in the same direction. If m_1 is less than m_2, then v_1' becomes negative.

v_2' (the velocity of the target cart, m_2, after the collision) can be similarly predicted by substituting $v_1' = v_1^2 - v_1$ in the conservation of momentum equation.

That is, if

$$m_1v_1 = m_1v_1' + m_2v_2'$$

then

$$m_1v_1 = m_1v_2' - m_1v_1 + m_2v_2'$$

and

$$2m_1v_1 = m_1v_2' + m_2v_2'$$

$$v_2' = \frac{2m_1v_1}{m_1 + m_2}$$

Here again, it is possible to predict by inspection of the derived equation that v_2', the velocity of m_2 (the target cart) after the collision, will be equal to v_1 before the collision if $m_1 = m_2$. If m_1 is greater than m_2 then v_2' would be greater than v_1. That is, the target cart would go off with greater velocity than m_1 came in with. If m_2 is larger than m_1, then v_2' will be smaller than v_1 but will still be in the positive direction.

Another demonstration which amazes most students, but which can be explained by the predictions above, is the hanging balls demonstration. About ten steel balls are hung side by side from long strings so that they just touch each other when hanging at rest. (See Figure 12-1.)

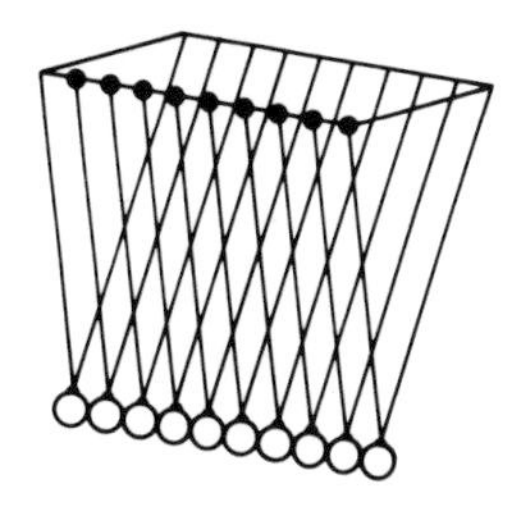

Figure 12-1

When a ball is lifted at one end and allowed to fall and strike the other balls, one ball is seen to rebound at the other end. When two balls are lifted at one end, two balls are seen to react at the other end, three, then three, four, then four, and so forth.

It can be argued that by momentum conservation alone, two balls coming down may cause one ball to go off at twice the speed, but then energy conservation laws would not be satisfied.

In order to satisfy the laws of conservation of both momentum and energy, the apparatus does not have any choice or decision to make. It must behave as demonstrated.

Of course, one should point out that it is so because the steel balls are extremely elastic. If these were not elastic collisions, then the conservation of kinetic energy would not be one of the controlling factors.

A most useful laboratory exercise for students is an extension of the PSSC laboratory on collisions in two dimensions.

The experiment is originally designed to show that momentum is conserved vectorially.

That is, if a steel ball rolling down an incline strikes another steel ball of equal mass at some collision other than head-on, the two will go off at such a path and with the proper velocity so that the vector sum of the momenta can be shown to equal the original momentum of the moving ball before the collision. (See Figure 12-2.)

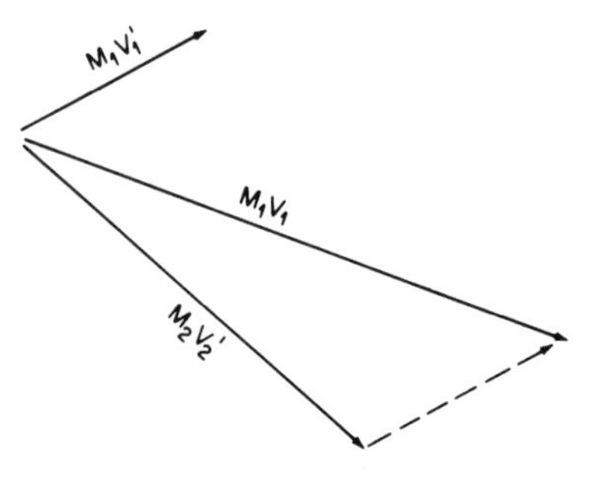

Figure 12-2

Momentum then is also shown to be conserved even when the target ball is less than the projectile ball in that the velocity of the target ball would be three times as great if the mass of the target ball is one third that of the projectile.

However, if we stick to colliding balls of equal mass, we can ask our students to look at the angle of departure after the collision and observe that these angles are never greater than 90°. To explain this, they will have to bear in mind that momentum is always conserved vectorially, but kinetic energy is conserved only in perfectly elastic collisions and energy is a scalar quantity.

Arguments which predict the observation proceed as follows (see Figure 12-3):

$m_1\vec{v}_1 = m_2\vec{v}'_2 + m_1\vec{v}'_1$. This is predicted by conservation of momentum and will also be shown to be confirmed experimentally. If $m_1 = m_2$, then

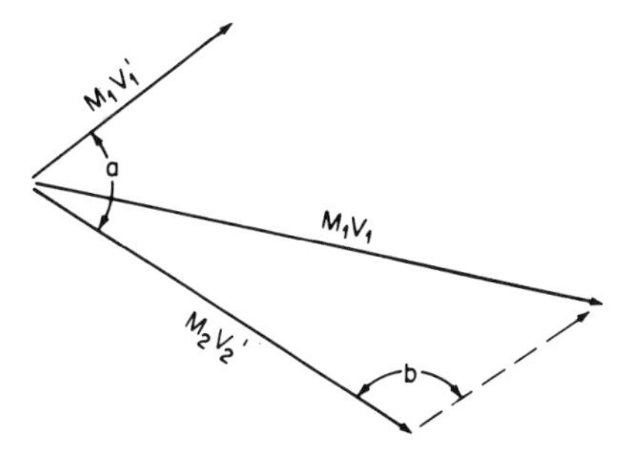

Figure 12-3

$\vec{v}_1 = \vec{v}_1' + \vec{v}_2'$

If kinetic energy is also conserved, then

$\frac{1}{2} m_1 v_1^2 = \frac{1}{2} m_1 v_1'^2 + \frac{1}{2} m_2 v_2'^2$

If again

$m_1 = m_2, v_1^2 = v_1'^2 + v_2'^2$

In order for $\vec{v}_1$ to equal $\vec{v}_1' + \vec{v}_2'$ vectorially and v_1^2 to equal $v_1'^2 + v_2'^2$ scalarly, a pythagorean relationship must be set up and angle (*b*) in Figure 12-3 must be a right angle. Angle (*a*) must also be a right angle, and one can safely predict that if two equal masses collide in an elastic collision other than head-on, they will always depart from each other at right angles.

If the collision is not elastic; that is, if some kinetic energy is lost through deformation (a soft collision) or by imparting spin, then angle (*b*) must be larger than 90°. Inspection of Figure 12-3 shows that when angle (*b*) is less than 90°, $v_1'^2 + v_2'^2$ would be greater than v_1^2 and more kinetic energy would be present after the collision than before. If (*b*) is larger than 90°, then angle (*a*) must be smaller than a right angle and one can conclude that in collisions between two objects of equal mass (other than head-on), the angle of departure can only approach a right angle but never exceed it.

Having demonstrated potential energy in bringing together two magnets with repelling fields, reverse the polarity of one of the magnets and show that work has to be done to pull them apart, and when they are held apart there is no apparent energy present (the energy is now stored in the field as potential energy). When the magnets are released, their energy is again visible as kinetic energy.

The notion of doing work when pulling attracting bodies away from each other is now extended to raising a mass away from the surface of the earth. The product of the force or weight (*mg*) and the height (*h*) is the amount of work needed to pull the earth and the object from each other. Releasing the weight is seen to put the kinetic energy back into the system as in the case of the magnets above.

The work (calculated as force times distance) is usually easily assimilated by students when the force is constant, as it is near the surface of the earth. For the measurement of potential energy, when the distances become large enough for the inverse square law to be effective, they will need a little more assistance from the teacher.

Two arguments regarding the uniform force field near the surface of the earth follow:

a. Intensity of a force field may be measured by the number of lines of force per unit cross-sectional area. Since gravitational lines of force are perpendicular to the surface of the earth, as long as you are close enough to the surface so that surface approximates a flat plane, gravitational lines of force are parallel and field is uniform.

b. Gravitational field strength is inversely proportional to the square of the distance from center of earth. A short distance from surface would be insignificant compared to the radius of earth (distance from surface to center). The short distance an object is raised near the earth's surface may, therefore, be thought of as being raised through a constant field.

To illustrate the derivation of potential energy with an inverse square law force, the technique of measuring the area under a graph is most effective.

If one does work against a constant force, the graph of force plotted against distance looks like the graph in Figure 12-4.

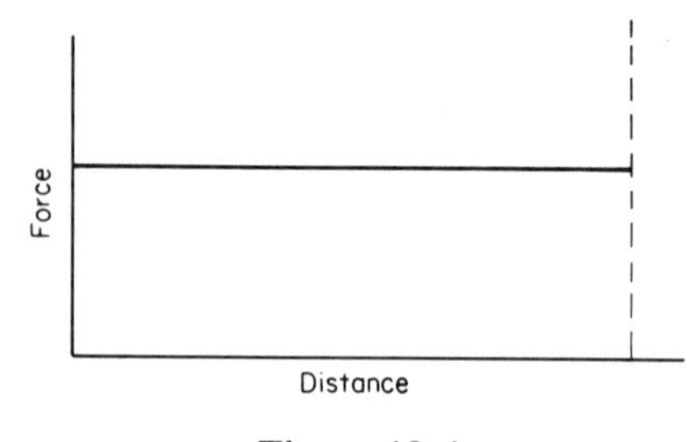

Figure 12-4

The work done is the force x distance or the area of the rectangle under the graph.

If however, the force is an inverse square law force, the graph of force plotted against distance would look as in Figure 12-5.

The force falls off to zero as an asymptote as the distance approaches

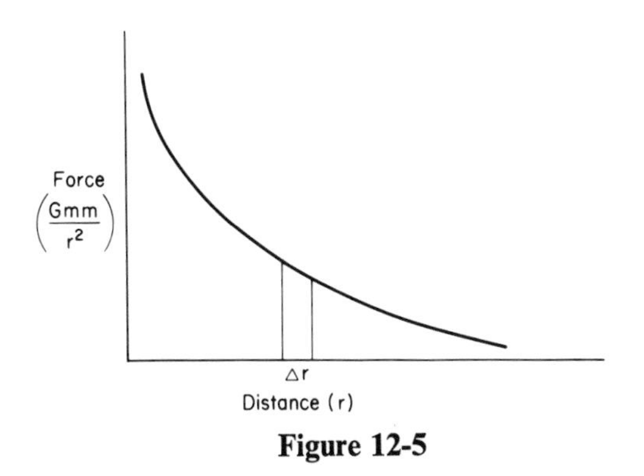

Figure 12-5

infinity. If the force is a gravitational force, then the total area under the graph is the summation of all the little rectangles ($Gmm/r^2 \cdot \Delta r$), where the Δr's are very small increments of distance. One can see that by integrating $\int Gmm \cdot r^{-2}\, dr = -\,Gmm/r$.

Your students have probably not had calculus, but you should be able to point out the necessity for using a mathematics which permits the summed effects of continuous change and hopefully convince them of the rationale of what you are doing, even if they do not understand all the mathematics.

The concept of minus energy is troublesome at first, but once the student grasps that energy content is relative to some reference point, it makes it easier. For instance, the zero point potential in the gravitational situation would be reached at some infinite distance from the attracting source. This concept is most useful when applied to the atomic situation in which the minus potential measures the amount of work necessary for ionization, that is, the "work function." Any energy supplied to an electron in excess of its potential relative to the ion from which it is being removed would then go into increasing its kinetic energy.

Another helpful idea is to adopt the convention that attractive forces are negative and repelling forces are positive. Thus the direction of gravitational forces is negative (we had already adopted the convention of calling the downward direction negative). This will fit in later in the case of Coulomb's Law where forces between unlike charges are attractive and the product of a positive and negative charge is negative. The repulsive force between like charges is proportional to the product of these charges, and the product of plus and plus or minus and minus is always positive.

POTENTIAL ENERGY STORED IN SPRINGS

The potential energy stored when work is done in compressing or stretching Hooke's Law type of elastic medium like a spring is very easy to calculate.

Plotting the force against the amount of extension or compression of a spring, one gets a graph which illustrates that the force is proportional to the amount of compression or elongation. (See Figure 12-6.)

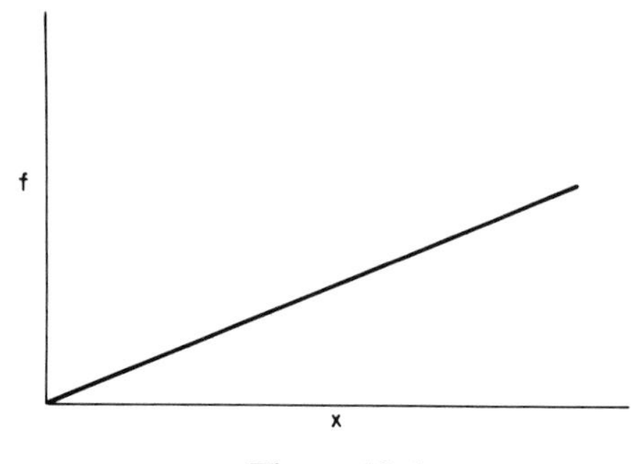

Figure 12-6

Since the area of triangle under the graph is $\frac{1}{2} F \cdot x$ and $F = kx$, then the potential energy stored in a spring is $\frac{1}{2} kx^2$.

A very good experiment for emphasizing the conservation of energy and transferability of energy from one form to another is the one in which a weight is hung from a spring and allowed to oscillate up and down.

A strong caution required here is that most students do not recognize the residual $\frac{1}{2} kx^2$ left in the spring at the top of its rise. I found this to be true also at summer institutes for teachers, where teachers experienced this laboratory exercise for the first time.

A spring is hung from a support and its elongation is measured for several weights. The student is then asked to plot the elongation against the causative weight to determine that the spring is truly obeying Hooke's Law and also to determine the constant for that particular spring (see Figure 12-6). He is then asked to raise the weight or lower it, and then release it so that it now oscillates up and down. (The extra energy given must be less than the amount already given by the hanging weight so that some residual stretch is left at the top of the oscillation, otherwise the spring will be either compressed or will lose energy in an unmeasured fashion as its coils collide with each other and it bounces around at the top of its rise.) (See Figure 12-7.)

The student observes that the weight oscillates between two points equidistant from the rest position where $kx = -mg$. It should be pointed out here that the experiment is only feasible because the mass of the spring is insignificant compared to the mass of the oscillating weight.

All that remains is to observe that three forms of energy are now manifesting themselves—kinetic energy ($\frac{1}{2} mv^2$), potential energy in the spring ($\frac{1}{2} kx^2$), and gravitational potential (mgh).

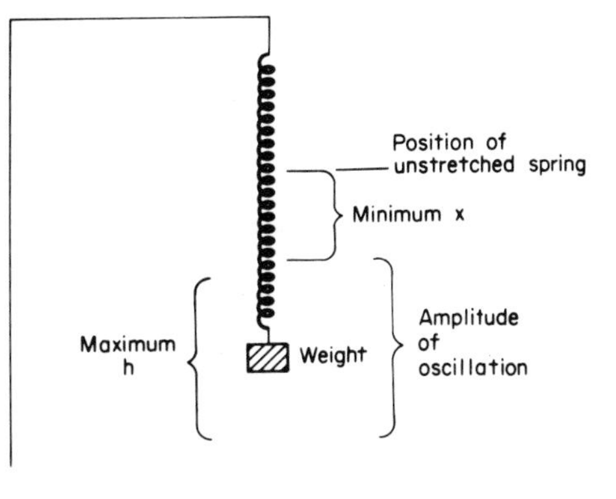

Figure 12-7

By plotting *mgh* vs. x, $\frac{1}{2}kx^2$ vs. x, and the total potential $mgh + \frac{1}{2}kx^2$ vs. x, graphs are constructed and energy conservation comes to light even though there is a constant transfer from one form of energy to another.

An example of such an operation follows:

Assume a spring, whose $k = 10$ newtons/meters, has a 1 kg mass hung on the end of the spring stretched at one meter. Stretch it another half meter and release the weight so that it oscillates through an amplitude of one meter from an elongation of a half meter to one and one-half meters.

Construct a table of data for various *mgh* at different x's (h is always measured from the bottom of the oscillation) and for various $\frac{1}{2}kx^2$ for different x's (x is always measured from the point where the bottom of the unstretched spring would be if no weight were hanging from it). (See Figure 12-8.)

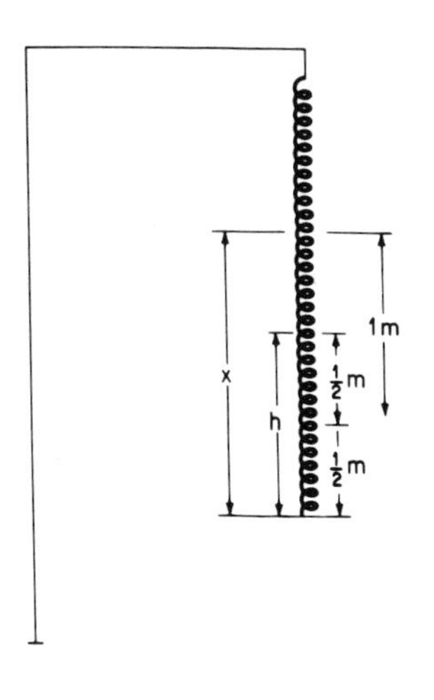

Figure 12-8

x	mgh	$\frac{1}{2}kx^2$	$U_t = mgh + \frac{1}{2}kx^2$
.5	10 newton meters	1.25 joules	11.25
.6	9 newton meters	1.80 joules	10.80
.7	8 newton meters	2.454 joules	10.45
.8	7 newton meters	3.20 joules	10.20
.9	6 newton meters	4.05 joules	10.05
1.0	5 newton meters	5.0 joules	10.00
1.1	4 newton meters	6.05 joules	10.05
1.2	3 newton meters	7.20 joules	10.20
1.3	2 newton meters	8.44 joules	10.44
1.4	1 newton meters	9.8 joules	10.80
1.5	0 newton meters	11.25 joules	11.25

Before graphing, one can see a symmetry in the data of u_t vs. x. The potential is the same at the top and bottom and reaches a maximum there while the combined potential is minimal in the center of the swing. One ought to remember that in our previous examination of the simple harmonic motion exhibited by an oscillating spring, the minimum velocity was reached at the two ends of a half cycle while the maximum velocity occurred in the center of the half cycle. Your students ought to be able to recognize the confirmation of this in the data just accumulated. (If energy is conserved, then maximum kinetic would be evidenced where you have minimum potential.) The graphs would look like Figures 12-9 and 12-10.

With the completed graph in his possession, the student can now be asked to determine the velocity of the mass as it goes by any point in its oscillatory cycles. Knowing that the kinetic energy is zero at the top and bottom of the cycle, he should recognize that the difference from the total potential at any point of the cycle and the potential at the ends is the kinetic. If he knows the kinetic energy and the mass of the moving weight, he should be able to calculate the speed of the weight.

POWER

Power defines the rate at which one does work or uses energy. The unit of power, the watt or joule/second, is probably the notation most familiar to your students, although for the car enthusiast, horse power is still a familiar term. A confusing bit of terminology for some is the local power and light company which sells "electric power" and bills them for kilowatt-hours. It should be shown (by dimensional analysis), that a kilowatt-hour is a unit of energy, not power, and that the light company sells energy, not power. However, if it is a powerful light company, it can supply a great deal of energy per unit time.

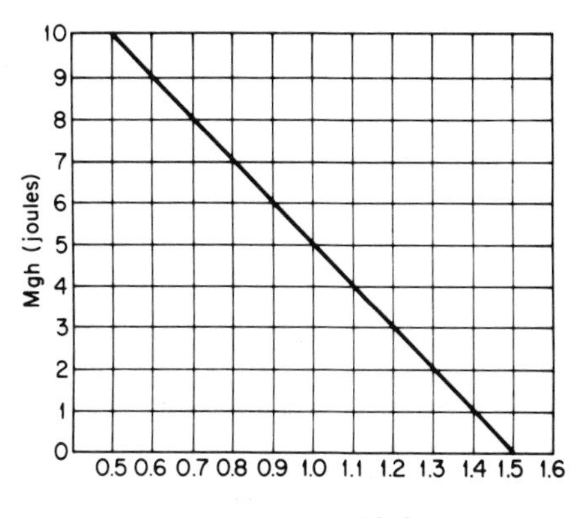

Figure 12-9

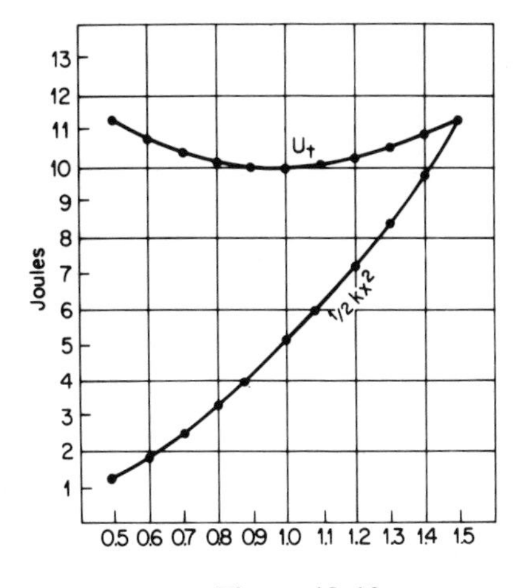

Figure 12-10

13

Gas Model

IN CHAPTER 3 WE DERIVED or, at least, rationalized the ideal gas law $pv = n_0Nkt$. It is interesting to note that pv (pressure times volume) has the units of energy. That is, newtons/meter^2 $\times$ meters^3 = newton meters or joules. Since temperature is a measure of the kinetic energy of the particles making up our gas and the Boltzman constant (k) in the $m \cdot k \cdot s$ system has units of joules/°K, the equation for the ideal gas law is dimensionally correct. Now that your students have a little better understanding of what is meant by energy and work, it is a good time to reexamine the gas model built up earlier from the standpoint of energy.

Experimental Evidence

We usually cite the "Joule" experiment in which the potential energy lost by dropping weights is converted to thermal energy as the temperature of a tank of water rises because of the rotating paddle in the water. An experiment performed about the same time as Joule's work, arriving at the same results, was described by Julius Meyer and emphasizes the energy relationships in the gas model.

Visualize a well-insulated jacketed cylinder with a piston having one side exposed to the atmosphere. (See Figure 13-1.) Hot water is pumped through the jacket and the temperature of the water input and output is recorded. If we know the water flow and temperature differential, we know the heat flow. For instance, if 100 cc of water per minute is pumped through the jacket and the input water is 90°C while the output is 85°C, then 500 calories per minute are transferred to the air inside the cylinder.

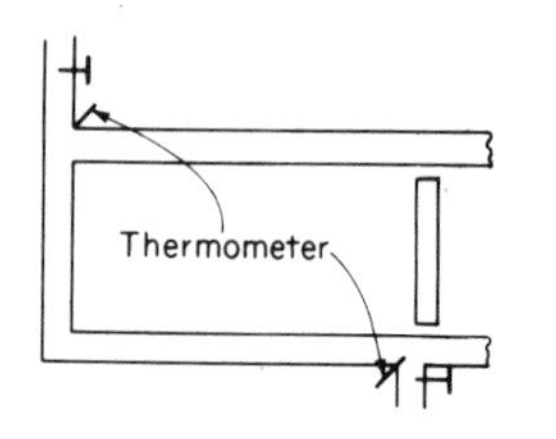

Figure 13-1

The pressure holding the piston in is one atmosphere and the pressure inside moves the piston out as the gas inside the cylinder expands until the pressure inside the cylinder is also one atmosphere.

In order for the piston to move out, work had to be done since the pressure of the air on the outer face of the piston (1.01×10^5 newtons/meter^2) times the area of the face must equal the force holding the piston in, and thus this force times the distance the piston moves out in order to permit the gas to expand must represent the work done on the piston by the expanding gas.

$$p = \frac{f}{a} \text{ (by definition)}$$

$$\frac{f}{a}\, a = \text{force}$$

$$f \Delta x = \text{work}$$

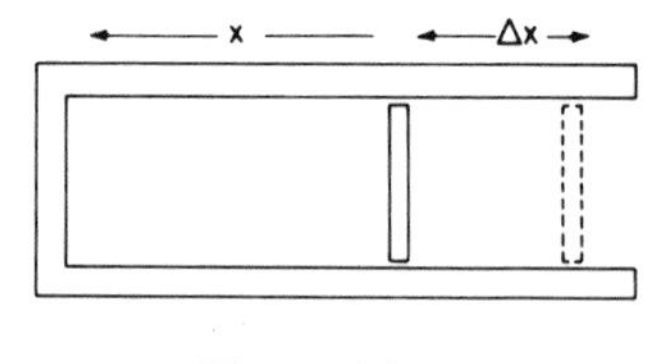

Figure 13-2

$$a \text{ (area of piston) } \Delta x = \Delta \text{ volume}$$

$$p \cdot \Delta V = \frac{f}{a} \cdot a \cdot \Delta x = f \cdot \Delta x$$

This $f\,\Delta x$ calculated in joules and then equated to the number of calories previously observed being transferred to the gas by the circulating hot water gives the same relationship Joules found in his conversion of gravitational potential to heat. That is, approximately 4.2 joules = 1 calorie.

Incidentally, it ought to be pointed out to your students that the calorie referred to here is the small calorie required to raise 1 gm of water 1° centigrade. The calorie one does not eat in order to lose weight is a kilo cal (1000 of these small calories).

Temperature and Energy

As each particle of gas with momentum mv strikes the wall of the container and rebounds with equal and opposite mv, it exerts a force on the wall equal to $2mv/t$. Considering the total force exerted by n particles, the wall would be subjected to a force of $2mvn/t$ if all the particles were moving perpendicular to the wall. However, considering the motion of all the particles perpendicular to the wall, half are moving toward the wall and half are moving away from the wall, so the force is then mvn/t.

You can then assume that the force exerted on any unit area is due to the impacts of all the particles in the volume of an imaginary cylinder perpendicular to the wall. (See Figure 13-3.)

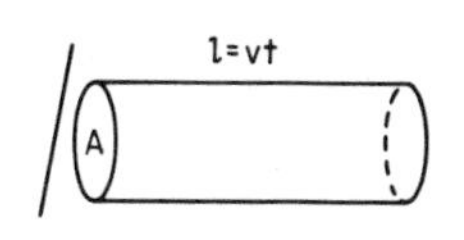

Figure 13-3

The volume of this cylinder is $a \cdot vt$

The number per volume $\frac{(n)}{V}$ is $\frac{n}{a \cdot vt}$

and $n = \frac{n}{V} \times a \cdot vt$

since $f = \frac{mvn}{t}$

$$f = \frac{mv}{t} \cdot \frac{n}{V} \cdot a \cdot vt$$

and the pressure is $\frac{f}{a} = \frac{mv}{t} \cdot \frac{n}{V} \cdot vt,$

$$p = \frac{mv^2}{V} \cdot n$$

so $pV = n \cdot mv^2$

a is area
v is velocity
t is time
V is volume
m is mass
f is force
p is pressure
n is number of particles

The above assumes that all the motion is back and forth on a line perpendicular to the wall on which the force being measured is exerted. If we call this line of motion the x axis, we can assume the possibility of motion in three-dimensional space. The y and z axis are both perpendicular to the x axis and parallel to the wall. Since all three modes of motion are equally possible and the population of moving gas particles is very high, then the chance of all possible modes being equally populated becomes most probable. (See Figure 13-4.)

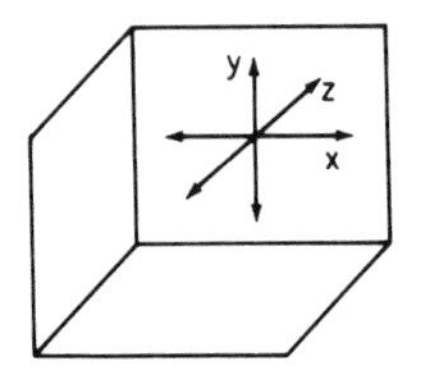

Figure 13-4

The particles moving in the y and z planes do not contribute to the impacts on the wall (both motions are parallel to the wall). The pressure, therefore, is:

$$p = \frac{1}{3}\frac{mv^2}{V} \cdot n$$

or $pV = \frac{1}{3} mv^2 \cdot n$

$$mv^2 = 2e_k$$

and $pV = \frac{2}{3} e_k \cdot n$

The kinetic energy (e_k) is, of course, the average kinetic energy of each particle, and $e_k \cdot n$ is the total kinetic energy contained by this gas sample.

No prediction can be made about the motion or energy of any single particle, but on a statistical basis the total energy in this sample of monomolecular gas can be safely predicted to be

$$e_k = \frac{3}{2} p \cdot V$$

Since $pV = nNkt$ (where $n =$ no. of moles and N is Avogadro's number), then $\frac{2}{3} e_k = nNkt$ and the temperature in °k is directly proportional to the average kinetic energy of gas particles. The factor $\frac{1}{3}$ is true of monoatomic gases since there are three possible modes of energy absorption and only one mode is measurable by the pressure. In diatomic gases, energy can be absorbed in other modes. For instance, the two parts of a diatomic molecule can rotate around their common center of mass, they can vibrate back and forth exchanging potential and kinetic energy as energy is put into and taken out of the attracting field between them. Triatomic molecules have even more modes of energy containment.

We have been describing this motion as random kinetic energy. This is commonly referred to as thermal energy. Since energy is a scalar, the randomness does not affect the summation for measuring the total energy containment of a given gas sample.

Since this is random motion, the sum of all these must be perpendicular to the walls. For instance, any motion in the following direction ↘ with $+x$ and $-y$ directions is matched by another ↗ with $+x$ and $+y$ directions.

A Student Experiment

An interesting student valuation of the Joule experiment illustrating the transfer of gravitational potential to thermal energy can be set up as follows:

Tubes one meter long are filled with water. Lead shot of known mass is placed in the tubes and the tube is sealed with a one-holed stopper in which a thermometer is placed.

The tube is then turned end over end and the temperature differential of the water in the tube is noted. The potential energy in joules is compared to the calories accumulated by the water in raising its temperature. The equivalence of approximately 4.16 joules per calorie should be experimentally derived.

This is a good experiment because it does work as expected (by most students).

We put about 1000 gms. of lead in the tube containing about 225 cc of water. Each time the lead drops 32 cm (one meter) it loses 3.13 joules of potential energy. It would have to drop 4 times to lose one joule and about 16 times to lose one calorie. With 225 grams of water in the tube to warm up, it would take about 300 turns to heat the water up 1° centigrade. I usually have the tube set up, give the student the data, have them calculate the number of turns required, and then have them pass the tube around with each student giving it approximately ten turns until it gets the necessary 300 turns. The 1° centigrade rise is then duly noted and we have confirmed Joule's results.

The concept of sound transmission as a compression wave can help visualize the relationship between energy and temperature. The average free path of an air molecule at atmospheric pressure is about 300 molecular diameters. If increasing the temperature of a gas increases the kinetic energy or average velocity of gas particles, then it would predict that sound waves would travel faster in air when the temperature is higher. The fact that it is experimentally so is further verification.

A Demonstration

An interesting demonstration can be set up by putting about 2 cu.cm. of mercury into a Pyrex test tube. Float small pieces of wood chips on the mercury. Evacuate the test tube and with the test tube connected to the running vaccum pump, heat the mercury slowly. Soon the wood chips will begin to dance about and the higher the temperature, the more violently the wood chips will bounce about. A bright point source of light will project shadows of the dancing chips on a screen so that it becomes an easily seen demonstration.

14

Basic Static Electricity and Coulomb's Law

THE INTRODUCTION and/or review of static electricity lends itself to some interesting demonstrations for students.

Demonstration of Charge

The plastic strips of vinylite and cellulose acetate currently available from laboratory supply houses are excellent for holding charges. They can be charged much more easily than the glass and ebony rods we once used. According to instructions, the vinylite is supposed to take a negative charge when rubbed with wool, while the acetate takes a positive charge when rubbed with cotton. I have found that they will actually take opposite charges when you rub both with the same material. To indicate that there are two different kinds of charges, tie a loop on each of the different strips (so that you can hang them up), stroke each on your coat or trousers and hang them up. To illustrate that they are charged, show that each will cause a small piece of paper to jump up and adhere to them. This also illustrates the tremendous magnitude of electrical forces versus gravitational forces since the electrical charge on this little piece of plastic is enough to overcome the gravitational pull of the mass of all the objects (including the entire earth) beneath the piece of paper.

With the two oppositely charged strips hanging freely from a rod or

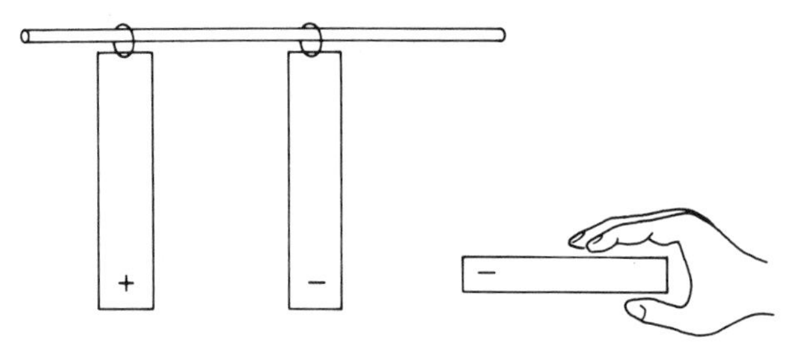

Figure 14-1

string, charge up a third strip, illustrate that it is charged by causing it to attract a piece of paper, and then show that it will attract one charged piece of plastic and repel the other. (See Figure 14-1.)

From here one can then continue with the usual demonstrations involving pith balls and electroscopes to show that like charges repel each other and unlike charges attract each other.

Charging electroscopes by induction and conduction and having your students figure out (with some help from you) why electroscopes charged by conduction bear the same charge as the charging medium and those charged by induction have the opposite charge is a very useful method of introducing charge flow in conductors.

Demonstration of Charge Motion

Another demonstration which can be used for demonstrating charge flow, charging capacitors, etc., can be set up as follows:

Place a pair of bent copper strips on beakers as in the sketch in Figure 14-2. Hang a pith ball between the two upright portions and bring the two upright copper strips close together just leaving enough room so that the pith ball can vibrate perceptibly between them. Charge a vinylite and acetate strip and hold them over the flat part of the copper strips on the beakers. (Do not touch the copper strips with the charged plastic.) (See Figure 14-3.)

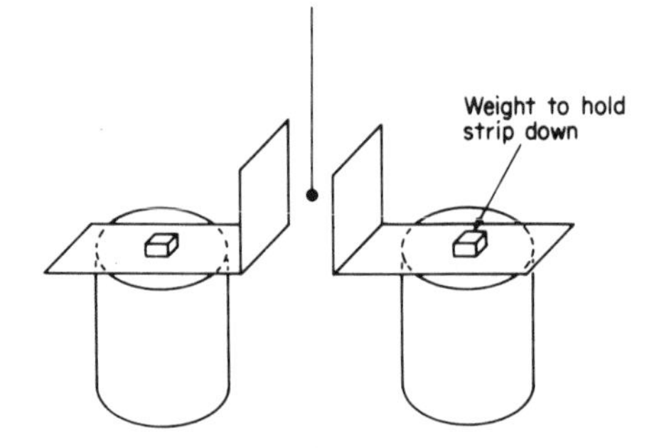

Figure 14-2

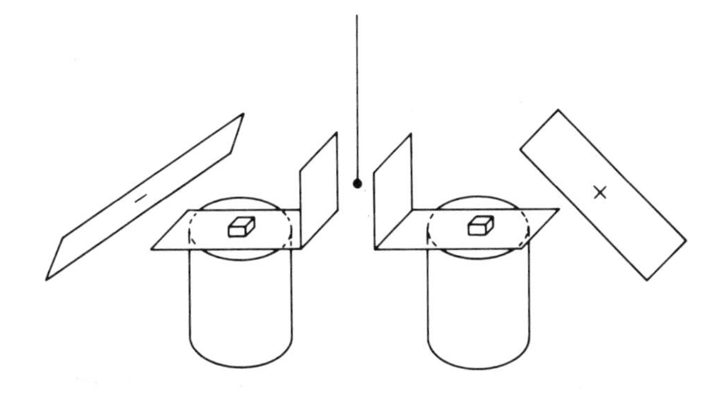

Figure 14-3

Observe the pith ball violently vibrating back and forth, transferring charge across the gap to compensate for the presence of the two charged plastics, come to a stop when equilibrium is reached and then when the plastic strips are removed begin to vibrate again as charge is retransferred to make up the old equilibrium.

Why the pith ball happens to go one way or the other at the beginning is purely random. It is probably just a little closer to one plate or the other, but once it goes over to one of the uprights it picks up the charge and is repelled to and attracted by the other plate where it unloads charge of one sign and picks up the opposite charge, thus being caused to go back and forth until the available charge on each plate is the same.

Electric Field

Having previously discussed the concept of field in the gravitational situation, it would now be appropriate to define and demonstrate electric fields. While field in the gravitational case was defined as force per unit mass as in newtons per kg, electric field is to be defined as force per unit charge as in newtons per coulomb.

Coulomb's Law Demonstration

One ought to first establish Coulomb's Law. This can be done as a class experiment as follows:

Place a transparent plastic sheet with rectilinear grid markings on the platform of your overhead projector. Hang a pith ball from a nylon monofilament over the grid and prepare another pith ball on the sharpened end of a piece of stiff piece of polyethylene which can then be held in a clothespin platform. (See Figure 14-4.)

The pith balls can be charged by rubbing against charged plastic strips

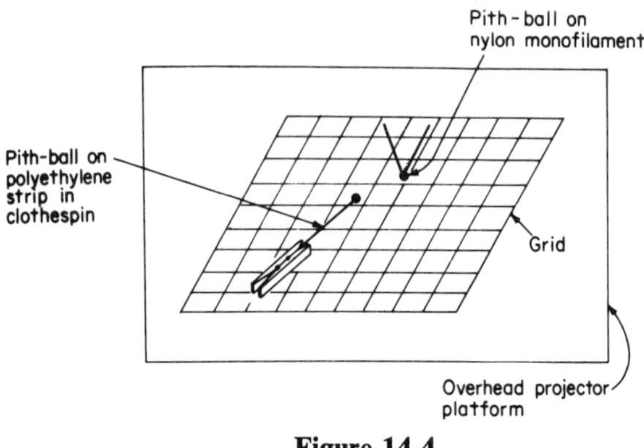

Figure 14-4

or, for best results, by transferring charge to them from a charged Van de Graff with a suitable capacitor.

When one is satisfied that both pith balls are suitably charged, the experiment can begin.

It is important that one first establish to everyone's satisfaction that the displacement from rest of the hanging pith ball is proportional to the force causing its displacement. A simple geometry is sufficient. (See Figure 14-5.)

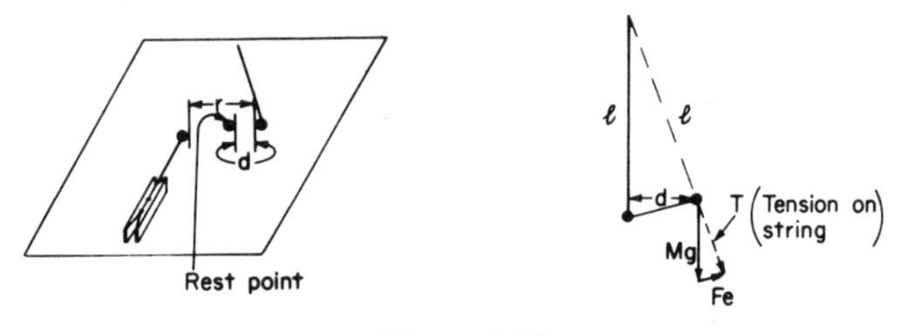

Figure 14-5

When the hanging charged ball is displaced due to the presence of another charge, the tension on the string can be resolved into two force vectors $\vec{m}_g$ and $\vec{f}_e$. Since the triangle m_g, T, f_e is similar to $\vec{1}$, $\vec{1}$, $\vec{d}$ then $1/m_g = d/f_e$. Since $1/m_g$ is a constant, $\vec{d}$ is proportional to $\vec{f}_e$.

With the projector turned on for class viewing and the charged pith ball on the clothespin an infinite distance away so that the ball on the string hangs freely under gravity alone, mark the position of one edge of its shadow on the grid with a crayon. Move the other ball in and note both the distance (r) between the shadow projections of the two balls and the resulting displacement (d) of the hanging half from its rest position when gravity alone was acting on it. Take data for at least ten different positions. Ask your students to find by graphical analysis a linear relationship between r and some function of d. They should come up with a reasonable proportionality of d vs. $1/r^2$ when r is not too small. As r gets very small this relationship falls off since the charges no longer see each other as point charges but rather as polarized accumulations of charge. Since we have already shown $d \alpha f$ then $f \alpha 1/r^2$. Now when either pith ball is touched with a equal sized uncharged pith ball (to remove half the charge) and the experiment is repeated, it will be seen that the slope of the f vs. $1/r^2$ graph is halved. When the other charged pith ball is made to share its charge with another uncharged pith ball and the experiment is repeated, the slope of f vs. $1/r^2$ is seen to be halved again. It then becomes evident that Coulomb's Law $f = k\,qq^2/r^2$ is a fairly accurate conclusion of the experimentally derived data.

It may be that because of atmospheric humidity, charges leak off the test spheres spoiling quantitative results. I have found that the presence of a charged Van de Graff machine (used for transferring charge to the test sphere) helps keep the atmosphere saturated with the right charge and minimizes leakage.

Demonstration of Vector Nature of Electrical Force

With the same apparatus mounted on the overhead projector, but with two movable pith balls equally charged (by sharing) acting on the hanging charged ball, show that the displacement of the hanging ball from its rest position under gravity alone can be predicted by summing up the force vectors acting on the hanging ball. (See Figure 14-6.)

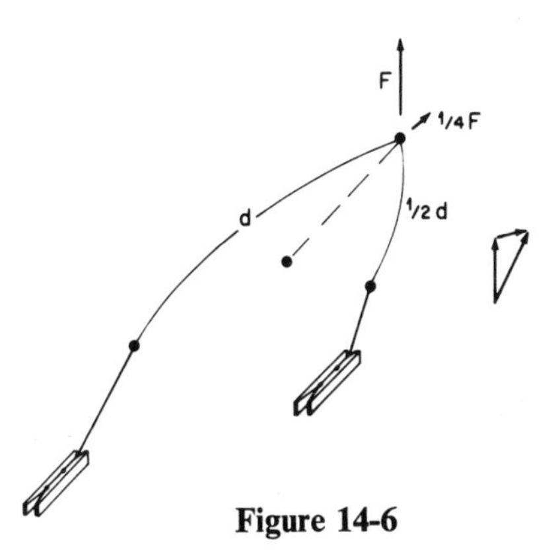

Figure 14-6

Demonstration of Force Fields

A most convenient method of demonstrating electric force fields is by using small grass seeds floating on carbon tetrachloride around electric charges.

Place a glass pan on top of the overhead projector platform. Standard sized pyrex pans used for baking are quite satisfactory. Put about one half inch of carbon tetrachloride into the pan. Dust a fine grass seed onto the surface of the carbon tetrachloride. There should be enough grass seed to give a fairly crowded look to the seeds floating on the liquid after it is stirred up (it will tend to clump when first dusted in) but with enough space between individual particles to insure free movement. Some experience will help determine the right amount. The grass seed recommended for this demonstration is a variety called "Colonial Bent" which is available at most Farmer's Co-ops and seed stores. Pour about one quarter inch of mineral oil on top of the carbon tetrachloride and stir up the grass seeds again as they will tend to clump again when the oil is poured in. I have found "Nugol" a very good brand of mineral oil for this purpose. The density of the grass seed lies between the carbon tetrachloride and the "Nugol" so that one can be assured that the grass seed will remain in the interface between the two liquids. The purpose of the mineral oil is to seal the carbon tetrachloride and keep it from vaporizing. Use a thin glass rod for stirring up the grass seeds. This will have to be done before each new demonstration. After you have set up the grass seeds and layers of liquid in the glass pan as above, you will probably be able to use it for all your classes all day. It will have to be reassembled fresh the next day if you are going to continue with such demonstrations since the grass seeds will soak up the carbon tetrachloride overnight and sink to the bottom of the pan.

Place two electrodes at each pole of a high voltage source. These electrodes should be stiff wires so that the ends can be safely immersed in the carbon tetrachloride on the pan of the overhead projector without danger of collapsing.

A neon transformer of about 8,000 → 14,000 volts is a satisfactory source. A Wimshurst generator also works very well. (The apparatus works equally well with A.C. or D.C. sources.)

First place one electrode in the tank and turn on the power. The seed will be seen to line up radially around the electrode (a point source).

Explanations to Student

While the seeds are non-conductors, they do become polarized in an electric field. Thus, they attract each other's plus and minus charges. If the source is strong, more will tend to line up on the source. If the source is weak, fewer will be moved toward it from a distance and there will, therefore, be fewer radiating lines around it. This can be demonstrated if the transformer is plugged into a variable A.C. (or by varying the speed of rotation of the Wimshurst). (See Figure 14-7.)

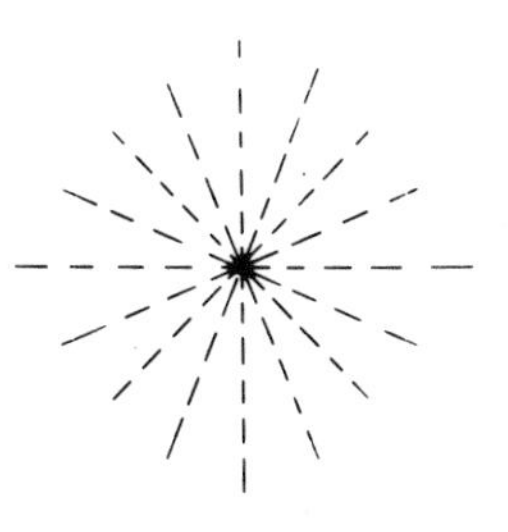

Figure 14-7

Putting the two electrodes into the tank and turning the current on will demonstrate lines of force between two point sources of opposite charge. Always remove the plug from the outlet before adjusting the electrodes as a good safety habit.

When the two electrodes are attached to the same side of the transformer or Wimshurst machine, the resulting pattern illustrates the field between two point charges of like sign. (See Figure 14-8.)

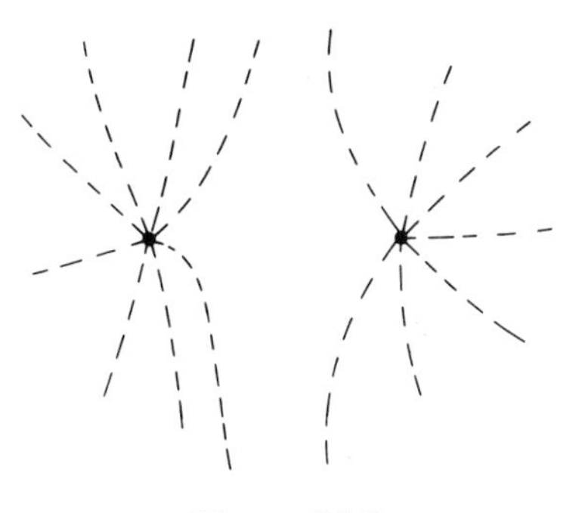

Figure 14-8

Point out that at a distance the field between the like charges appears to be perpendicular to a line connecting the charges.

Connect a copper strip to one of the electrodes and show (see Figure 14-9) that the field is always perpendicular to a line source (an infinite number of point charges of like sign).

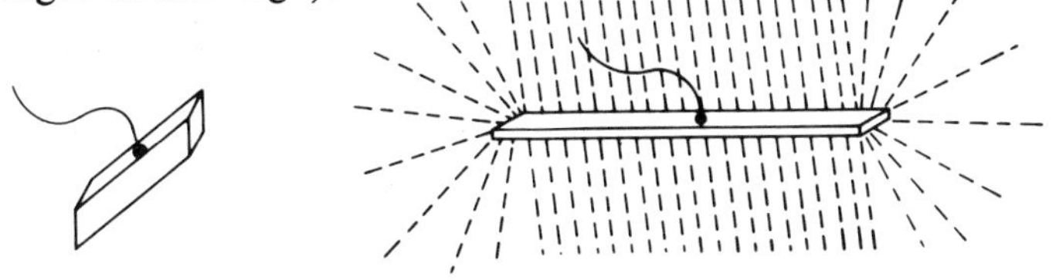

Figure 14-9

With two parallel line sources of opposite charge, the field lines, perpendicular to the two sources, are seen to be parallel to each other between the two sources. If the number of lines of force per cross-sectional area of field represents the strength of the field at that distance from the source, then the field between these two parallel line sources of opposite charge is seen to be uniform everywhere between them. Draw the analogy between this visible situation and the field between the plates of a capacitor.

Many other demonstrations might suggest themselves to you. One more that I find very useful can be described as follows: Obtain a piece of copper tubing about one inch high and $1\frac{1}{4}$ to $1\frac{1}{2}$ inches inside diameter. Close the ends with wide wire mesh. (See Figure 14-10.)

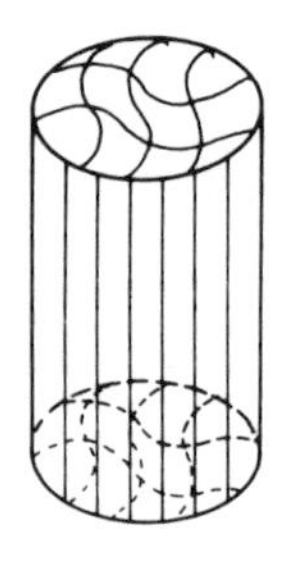

Figure 14-10

Place this tube on one end in the tank. Make sure that there are a good number of grass seeds floating on the surface of the carbon tetrachloride inside the hollow tube. Put one of the electrodes on the tube and take the other one outside the tank (disconnect it). Turn on the power and it will be observed that a radiating field exists outside the tube but no field is evidenced inside the hollow conductor, as shown by the random scattering of the grass seeds floating inside the tube. That this can be justified by Coulomb's Law can be demonstrated as follows (see Figure 14-11):

Assume a grass seed $\frac{1}{4}$ of the distance from one side of the conductor

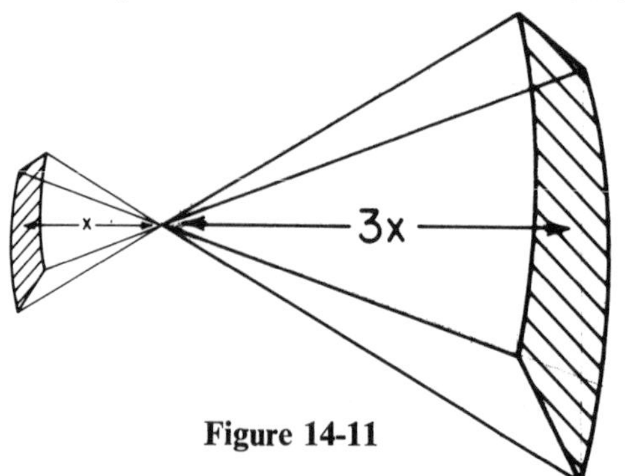

Figure 14-11

and $\frac{3}{4}$ from the opposite side. Visualize vertical solid angles to the two sides forming similar tetrahedrons with the base of one having nine times the area of the base of the smaller. The side with nine times as many charges exposed to the grass seed will be three times as far away, and if the inverse square relationship in Coulomb's Law holds, each charge will exert 1/9 the force on the grass seed as each charge in the closer side will. Since $9 \times 1/9 = 1$, the force is equal and opposite from the two walls. The charges in the grass seed will detect no field and the grass seed will not become polarized.

Another derivation of Coulomb's Law might be suggested by visualizing radiating lines of force around a point charge. If a point source radiates lines of force in three-dimensional space and the number of lines of force per unit area suggests the intensity of the field at a distance from that point source, then the intensity, expressed as $E = q_1/4\pi r^2$. q is constant for that point source (the number of charges in the source governing the number of lines of force originated from the source), $E =$ the field intensity and $4\pi r^2$ is the area of a sphere at a radius r from the square. Since E is defined as f/q^2, then $f = q_1 q_2/4\pi r^2$. This is usually expressed as $f = (1/4\pi\epsilon)\ (q_1 q_2/r^2)$ when ϵ is the permutivity of the space surrounding the charge which also affects the number of lines of force which radiate from it. ϵ for a vacuum is assigned the value one, and all other space has values of ϵ greater than one.

Since $1/4\pi\epsilon$ is a constant for the space being explored, then Coulomb's Law in the form previously derived $f = k\ (q_1 q_2)/r^2$ still holds.

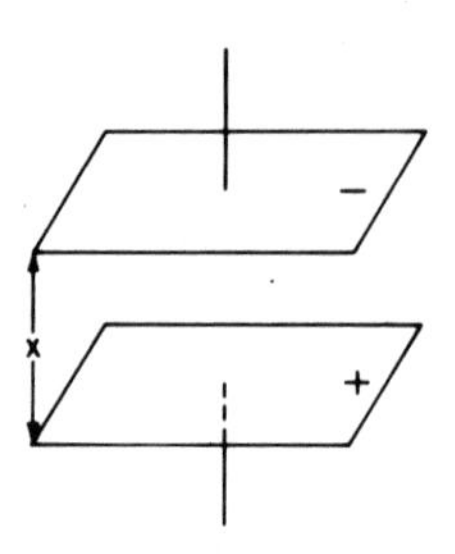

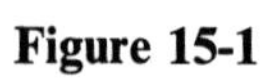
Figure 15-1

15

Electric Charge and Motion of Charge

Electrical charge instead of being spread uniformly over the charged surface has a definite granular structure, consisting, in fact, of an exact number of specks, or atoms of electricity, all precisely alike, peppered over the surface of the charged body.

ROBERT MILLIKAN IN 1911

AFTER A THROUGH STUDY of Newtonian mechanics, introducing electrical phenomena as another example of what has already been studied rather than as some isolated case has definite pedagogical advantages. One can point to the similarity in form of Newton's Law of universal gravitation ($F_g = Gm_1m_2/r^2$) and Coulomb's Law ($F_e = kq_1q_2/r^2$).

We can start with a charged capacitor. (See Figure 15-1.)

It had been previously shown in the grass seed demonstration that the field between charged plates of a capacitor is uniform. Therefore, a charged particle placed between the plates would experience a constant force and a uniform acceleration (assuming there was no friction). A positive charge placed near the plate in Figure 15-1 would experience an acceleration toward the negative plate and arrive with a kinetic energy $\frac{1}{2}mv^2$. If one now wished to put it back to its original position near the positive plate, one would have to do work $F \cdot x$ (x being the distance between the plates) equal to the $\frac{1}{2}mv^2$

it accumulated when it fell down the potential hill between the plates. This potential energy per charge is measured in volts, a volt being defined as a joule/coulomb.

Since volts (E/q) is equal to field (F/q) times distance (x), it follows that the field between the two plates of a capacitor must vary inversely as the distance between them if the potential across the two plates remains constant. On the other hand, if one keeps the plates a constant distance apart from each other, the field between the plates will vary directly with the potential across the plates.

How then does one generate a potential across the plates of a capacitor? Work must be done.

One uses some sort of electrical machine to transfer charge from one side of a capacitor to the other side. A battery which converts chemical energy to electrical energy, a device like a Wimshurst machine or a Van de Graff generator which converts mechanical energy to electrical potential, or a transformer which can convert the energy of A.C. currents into high electrical potential can all be used to pump charge up a potential hill and thus charge up a capacitor.

In our discussion of the motion of gravitational mass in a gravitational field, we defined the term terminal velocity, wherein a mass falling through an atmosphere collides with the particles of atmosphere along its fall. If the atmosphere was dense enough and the mass fell fast enough, it would collide enough times on the way down to give away as much kinetic energy as it accumulated by its loss in potential. It would thus fall at a constant rate.

The larger the falling mass, the more energy it would accumulate in falling since the field (F/m) times the distance it falls would equal the loss in potential. It would, therefore, have to strike more particles per unit time as it fell to achieve a terminal velocity than would a less massive object. This could be accomplished in two ways, by presenting a larger surface for collision or by falling faster. If its volume and, therefore, its surface remained constant (assuming a constant density and a spherical shape), then its terminal velocity would be proportional to its mass.

The electrical situation is exactly analogous. If charged spherical particles of constant mass and volume were placed in an electrical field containing an atmosphere, the particles would fall through the field at a constant rate depending on the quantity of charge in the particles. (The energy accumulated by falling and given away in collisions to maintain terminal velocity is equal to $\frac{F}{q} \cdot x$.)

Robert Millikan recognized this at the beginning of the twentieth century and used it in experiments to determine whether or not charge is continuous or occurs in some unit quantum. He reasoned that if one placed charged particles of controlled mass and density in an electrical field (while an atmosphere was present) in which gravity accelerated the mass while the electrical field also acted on the particle via its charge, he could determine the value of the charge by examining the terminal velocities of the particle.

He sprayed oil drops between the plates of a charged capacitor and observed their resultant motion through a microscope.

When the field was arranged so that the electrical force on a charged particle between the capacitor plates was upwards, he could then raise or lower the intensity of the field until the electrical force upwards on the particle exactly equalled the downward gravitational force. The particle then stood still. (See Figure 15-2.) Knowing the mass and, therefore, the weight of the particle, he thus knew the electrical force on this charged particle. Since the electric field between the plates is force/charge as in Newton/Coulomb, and the measured voltage across the plates is in units of Newton-meters/Coulomb, by dividing the potential in volts by the distance between the plates in meters (Newton-meters/(Coulomb/meters)), he could determine the value of the electric field between the plates.

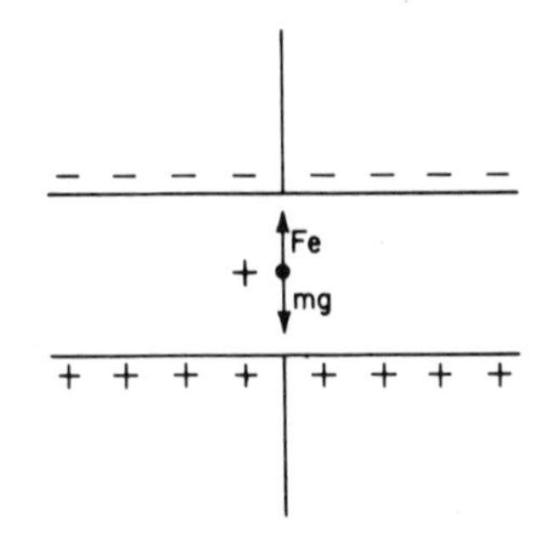

Figure 15-2

$$\text{Since Coulombs} \times \frac{\text{Newtons}}{\text{Coulomb}} = \text{kg} \times \frac{\text{Newtons}}{\text{kg}}$$

$$\text{or} \quad \text{Coulombs} = \frac{\text{Mass} \times \text{Gravitational Field}}{\text{Electrical Field}}.$$

It is then easy to calculate the amount of charge necessary to counterbalance this weight in this electrical field.

It can also be shown that if another charged particle could be seen to be moving upwards with the same velocity with which it would have fallen if there were no electric field between the plates, it must have exactly twice the number of charges that the balanced particle has on it. If another is observed to be falling with $1\frac{1}{2}$ times the velocity it would have, under gravity alone, it is because it has opposite charge of one-half the value of the previously determined balance charge. It was by such a technique that Millikan found, in his many observations of charged oil drops between the plates of his capacitor, that all measured charges were a small whole number multiple of a unit charge which came out to be about 1.602×10^{-19} Coulombs, or reciprocally, 1 coulomb equals 6.2425×10^{18} elementary charges. The important observation for students here is not the fact that an elementary charge on an electron or proton is equal to 1.602×10^{-19} coulombs—they can always look this up. What is important is the final proof of the fact that charge is grainy or quantized. This is fundamental to the understanding of atomic physics.

An Aside for Teachers

Millikan used oil drops in his classical experiments. There is available today a little plastic sphere of known mass and volume which is ideal for Millikan experiments in a high school laboratory. It might be interesting for you to know where these plastic spheres came from. They were not originally designed for high school Millikan apparatus, although they seem so ideally suited for the job. In the years that companies like Rohm & Haas, Dow Chemical and others were trying to develop a commercially useful ion-exchange resin, some experimental failures did occur.

Most commercial ion-exchange resins are made by polymerizing styrene monomer in the presence of a suitable peroxide catalyst with divinyl benzene for cross-linkage of the polystryene chains. The benzene rings contain the ion-exchange stations which are treated with sulfonic acids to produce cation exchangers or with tertiary amines for making anion-exchangers. The size of the polymerized beads formed is controlled by a number of parameters like the temperature of the polymerizing solution, the amount of catalyst present, and the speed of agitation.

An ion-exchange resin bead has an optimum size. If it is too large it will have too small a surface area and therefore not enough available ion-exchange surface for efficient operation of a practically sized column. If it is too small, it will increase the necessary pressure drop across the column to a practically inoperational point.

A batch of resin beads was made at Dow Chemical which were much too small for use as an ion-exchange resin. However, examination of these beads in an electron microscope showed them to be very uniform in size. In fact, for years batches of these beads were manufactured and sold for calibrating the field of electron microscopes. It was a sort of serendipity to find them available when the need for designing a high school type of Millikan experiment came up.

Capacitance

Note the apparatus suggested in Figure 14-2 to illustrate motion of charges. Now you can call attention to the high frequency of the vibrating pith ball at the beginning and the way it gradually slows down to a standstill. It then repeats the fast to slow vibrations and finally stops when the charged plastic is removed and the capacitor is allowed to discharge via the pith ball. The frequency of the pith ball is analogous to current flow. A good qualitative discussion of the properties of capacitors can follow this demonstration.

16

D. C. Electricity

IN A DISCUSSION of current electricity, one ought to first establish that the current flowing through an unbranched circuit is the same everywhere in the circuit. If the circuit contains branches, then the sum of the currents in the parallel branches must be the same as total current anywhere else in the circuit. This, of course, is summarized in Kirchoff's Laws but should be seen by students as a necessary and logical extrapolation of the basic laws of conservation.

For instance, after defining an ampere as one coulomb per second, or 6.25×10^{18} elementary charges per second, it can be argued that charges can flow only in a closed circuit; that is, from a place of high potential to a point of low potential. If a conductor were connected to one terminal of a D.C. power source, charge would flow into the conducting line accumulating at the other end until the reactance from the accumulated charge prevented any more from flowing. If the other end were grounded to the other terminal so that charge could flow off as fast as it accumulated, then a steady current would flow and the current would be the same anywhere.

This can be very easily demonstrated and probably should be. One need make up as complicated a circuit as one wishes with sufficient resistance to protect the power source and arrange it so that an ammeter can be placed in series anywhere in the line. You can show to the satisfaction of all that the current in the circuit is the same everywhere except in individual branches, but the sum of the currents in parallel branches is equal to the current flowing anywhere in the main circuit.

Direction of the Current

One of the minor controversies in the teaching of high school physics is over the convention of whether the direction of current should be the direction in which positive charges flow or in which negative charges flow. The fault for this is all Benjamin Franklin's since he, when assigning the sign of the charge given to a glass rod, after rubbing it with silk, arbitrarily called it positive.

Since then, textbooks on physics and electrical phenomena, definitive works like those of Maxwell and others have followed this convention and have established it as the convention. Remember, it is only a convention and changes nothing regarding physical reality. It is only since the use of metallic conductors in electrical technology and vacuum tubes in radio work, that some teachers and writers of high school textbooks have made such a fetish about changing the convention.

One need only remember that a negative charge moving in one direction is the exact equivalent of a positive charge moving in the opposite to realize that the controversy is hardly worth the effort. It has been my experience that once I have pointed this out to my students, they have no difficulty visualizing current flowing from the plate to the cathode of a diode even though it is physical electrons flowing from the cathode to the plate. I strongly advise teaching the proper convention as flow from positive to negative, since those who continue in physics on a more sophisticated level in college will find this to be the convention used and then will not have to unlearn negative flow, left-hand rules instead of right-hand rules, etc.

Actually, it seems to me the whole controversy is further minimized by the physical situation, since current carried by electron flow only, as in metal conductors or vacuum tubes, is truly a very specialized situation. In nature, currents are usually seen to exist in fluids (liquids, gases and plasmas) where both negative and positive ions flow, and the total current is determined by adding up all the ions moving by a point and assigning the direction as the direction of motion of the positive charges.

Ohm's Law

While the importance of Ohm's Law should not be overemphasized in the minds of your students, it is useful and can serve as a basic relationship for deriving other D.C. relationships. It ought to be first "discovered" experimentally in a student laboratory. The lab set-up is shown in Figure 16-1.

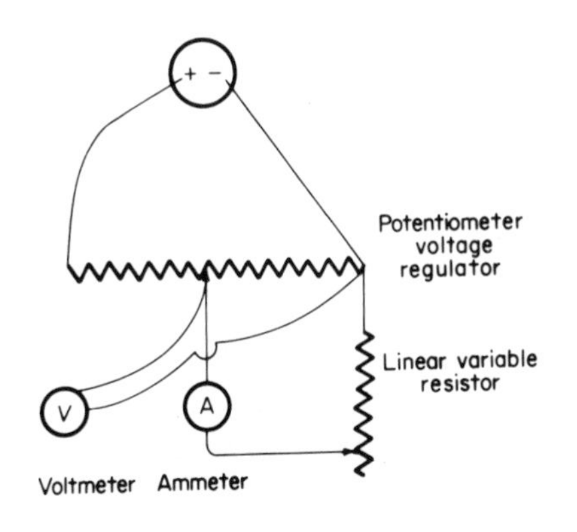

Figure 16-1

Keeping the resistance constant, students can record the current as they vary the voltage. With the voltage constant, the amperage vs. resistance (as in the length of the resistor) can be found. A simple graphing of the results will indicate that E (in volts) is proportional to I (current) and that I is inversely proportional to the length of the resistor (R).

It can, therefore, be deduced that $E = IR$ if the units of resistance are appropriate.

You can show with an ohm meter that the length of the resistor is proportional to its resistance.

It should be pointed out that the electric field in a conductor must be parallel to the sides of the conductor. This can be reasoned by the fact that charges moving in any direction other than down the length of the conductor from one terminal to the other would soon accumulate on the sides and then prevent further flow in that direction.

If the lines of force in the conductor are parallel, then that signifies that the field in the conductor is constant and the work done on a charge is simply $F/q \times x$. Therefore, the potential difference across the entire conductor is the sum of all the potential drops from one consecutive point along the line of current travel to the next. Another way of saying this is that when resistors are wired in series in a closed circuit, the potential drop across the entire circuit must equal the sum of the individual potential drops across each resistor ($E_t = E_1 + E_2 + E_3$, etc.).

The potential drop across the circuit must, of course, equal the e.m.f. supplied by the energy source. Incidentally, the term e.m.f. ought to be used in describing the source of the voltage supplied rather than the old term electromotive force. This is an ancient terminology built of some poor analogies. E.m.f. is a term used for the amount of energy supplied per unit charge. (Remember a volt is defined as a joule per coulomb.) Using the term electromotive force is confusing to beginning students and is reminiscent of a time when the relationship between force and energy was poorly understood.

If an analogy is useful, the proper analogy would be to the gravitational situation as follows (see Figure 16-2.):

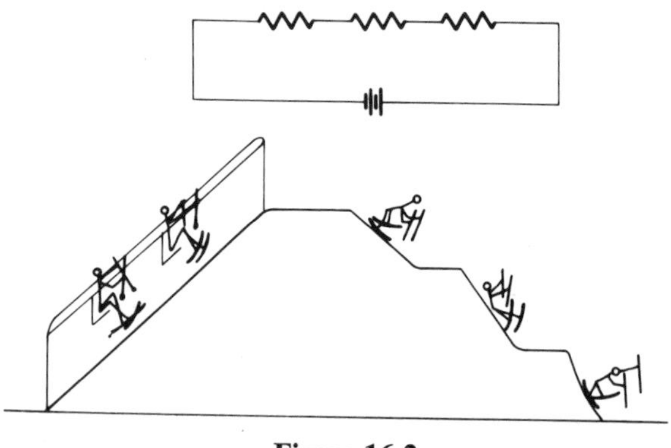

Figure 16-2

A group of resistors linked in series to a battery might be likened to a group of slopes down which skiers of equal mass come sliding after having been taken up to the top of the lift by the ski-lift (the e.m.f. machine) which is the source of their potential energy when they are at the top of the slopes. The downward slopes might be laden with obstacles like windblown branches of trees, etc., which keep getting in their way so that they must keep slowing down (applying friction) and the hill is so crowded they keep getting in each other's way. Of course, where the path of descent is narrower, fewer will get through per unit time (the resistance is higher and the current drops off).

The plateaus between each slope represent the conductors between the resistors where current can flow at a constant rate with negligible loss of potential energy. Resistors can be linked in parallel to the same battery. (See Figure 16-3.) The analogy with the skier would show them being lifted to the top of the potential hill with a choice of slope by which to descend. Thus each hill has a potential drop equal to the "e.m.f." supplied by the ski-lift and there is room for more skiers to come down at the same time. In a branched or parallel circuit, the potential drop across each resistor is the same and the currents in each resistor must add up to equal the current in the main line.

The laws of resistances in parallel and series can be deduced by using

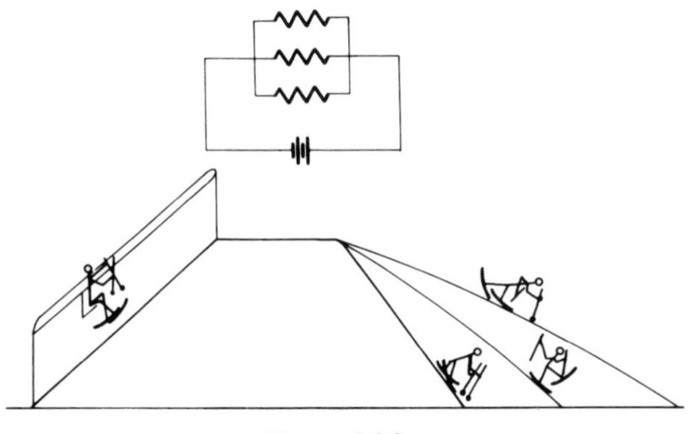

Figure 16-3

Ohm's Law and the fact that the voltage drop across resistors in parallel is the same for all the resistors in the branches and when resistors are arranged in series, the total voltage drop is the sum of the individual voltage drop across each resistor.

For instance, for resistors in series

$E_t = E_1 + E_2 + E_3$, etc., but $E = IR$ and the current (I) in each resistor in the series is the same so,

$IR_t = IR_1 + IR_2 + IR_3$

Since the current is the same, we may divide both sides of the equation by I and $R_t = R_1 + R_2 + R_3$.

When resistors are hooked up in parallel, the sum of the currents in each branch is equal to the total current flowing in the circuit.

That is $I_t = I_1 + I_2 + I_3$, but according to Ohm's Law $I = E/R$, and since $E_t = E_1 = E_2 = E_3$, then

$$\frac{E}{R_t} = \frac{E}{R_1} + \frac{E}{R_2} + \frac{E}{R_3}$$

and

$$\frac{1}{R_t} = \frac{1}{R_1} + \frac{1}{R_2} + \frac{1}{R_3}$$

All the above should, of course, be verified by demonstrations by the teacher or by student measurements of assembled circuits with the appropriate voltmeters, ammeters, and ohm meters.

The effect of batteries with cells in series or in parallel should also be explored both empirically by demonstration or by student measurements with appropriate instruments and by discussion of the theoretical aspects.

A simple discussion might ensue as follows (see Figure 16-4.):

Imagine a vessel filled with sulfuric acid solution. When zinc and copper plates are put into the solution, the more active zinc tends to go into solution in the sulphuric acid more rapidly than the copper does. When the zinc is in solution, it must be present as positive ions, and excess negative charge is, therefore, left on the zinc plate. Some of the copper might also go into solution as Cu^{++} ions, but it can easily be seen that the positive ions in solution (H^+, Zn^{++} and Cu^{++}) would soon inhibit any further activity.

Now connect the Zn and Cu electrodes with an external conductor. The excess negative charges on the Zn plate will now flow to the Cu plate through the conductor provided.

As negative ions are thus prevented from building up on the Zn electrode, more Zn^{++} ions can go into solution. (See Figure 16-5.) The potential

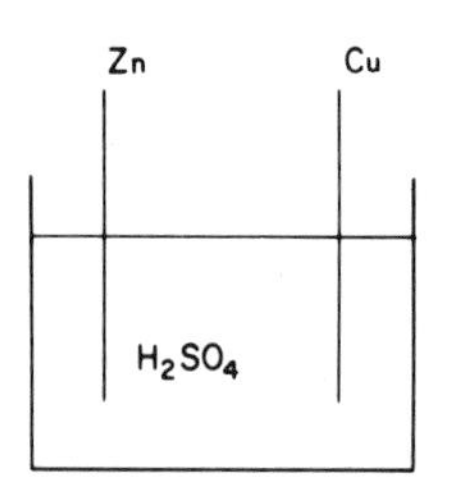

Figure 16-4

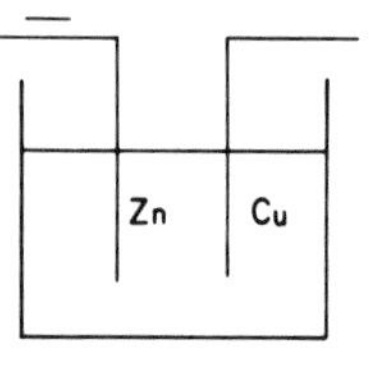

Figure 16-5

thus developed depends on the relative electrochemical activity of Zn and Cu.

The current will continue to flow as long as there is a closed circuit and the electrodes are not corroded beyond use. The cell would, of course, last longer if the electrodes were larger. Wiring several cells in parallel would be the same as having larger cells—the e.m.f. is not affected, the battery will just last longer.

When the cells are wired in series, each starts its potential ground level from where the preceeding one leaves off, and for cells in series, the total e.m.f. is the sum of the e.m.f's of the individual cells in the series.

Again, this should be checked and demonstrated with appropriate instruments on batteries of varying hook-ups. *Note:* What happens when cells of unequal e.m.f. are used in parallel?

It is useful in this context to illustrate the relationship between electrical energy and heat.

An electrical heating coil immersed in a known volume of water wired as in the diagram in Figure 16-6 is a typical student set-up.

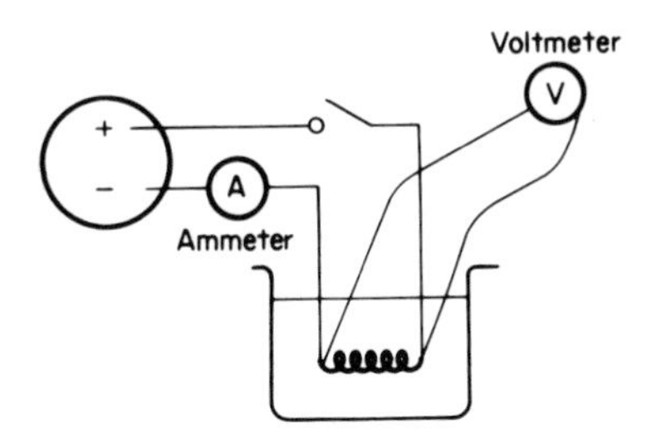

Figure 16-6

With the voltage and current known, the power in joules/sec. is known, the time the power is turned on times the power, gives the total energy fed to the water, and the calorie equivalent received by the water can be deduced by multiplying the number of grams of water in the beaker by the temperature rise in degrees centigrade.

The discussion following the student's laboratory experience can justify their findings.

$$\text{Amperes} \times \text{volts} = \text{watts}$$

$$\left(\frac{\text{coulombs}}{\text{second}} \times \frac{\text{joules}}{\text{coulomb}} = \text{joules/second}\right)$$

$$\text{joules/sec} \times \text{sec} = \text{joules}$$

$$4.2 \text{ joules} = 1 \text{ calorie}$$

$$1 \text{ calorie} = .24 \text{ joules}$$

$$\text{or} \quad \text{heat (in calories)} = .24 \text{ wt}$$

$$= .24 \text{ EIt}$$

but $E = IR$

and heat (in calories) $= .24\, I^2Rt$

Determination of Avogadro's Number

Another useful student experiment is to verify Avogadro's number by using the concepts just discussed.

Set up your apparatus as in the diagram in Figure 16-7.

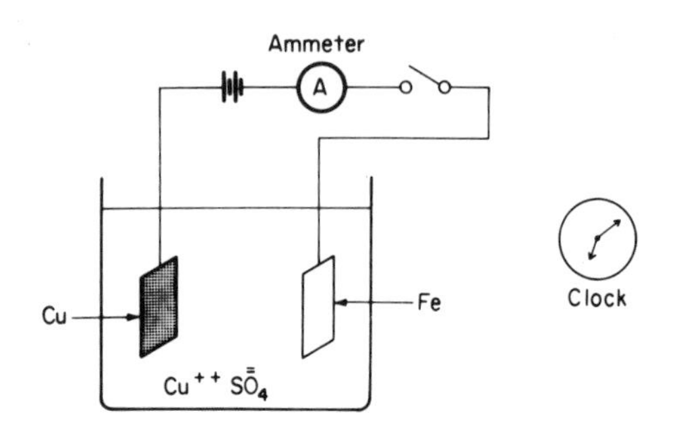

Figure 16-7

The iron electrode is dried (oven dried if possible), weighed, and then placed in the Cu SO_4 solution attached to the negative side of a D.C. source. Close the switch. Cu^{++} ions will then be attracted to the negative iron electrode, acquire two electrons per ion, become insoluble copper metal, and adhere to the metal cathode. Sulphate ions will gravitate to the copper electrode, solubilize the copper by combination as in $CuSO_4$, and thus bring Cu back into solution. The cycle continues as long as current flows in the circuit.

Record the amperage and the time the current is on. Dry the copper plated iron electrode and reweigh it.

One mole of Cu has a mass of 63.54 gms. The fraction of a mole is, therefore, determined.

The total flow of electrons can be calculated by multiplying the amperage (coulombs/sec) $\times$ time in seconds and multiplying the result by 6.25×10^{18} electrons/coulomb.

Since two electrons are necessary to sequester each doubly charged copper ion, one divides the number of electrons by two to determine the number of copper ions in the fraction of a mole collected on the iron electrode.

Copper atoms per mole can now be calculated by dividing the number of atoms collected by the fraction of a mole determined (atoms/mole). A reasonably close figure to Avogadro's Number should be attained. Even if the iron electrode were not air dried before and after plating (if an oven is not available), one ought to get a result within an order of magnitude of Avogadro's Number.

This experiment has several advantages. Besides illustrating movement of ions in solution when a current is flowing, it supplements the result previously attained in the oleic acid experiment.

Vacuum Tubes

An introduction to vacuum tube operation might be useful at this point.

It is easy to visualize that atoms in close proximity, as in a metal filament, would tend to bump each other in their thermal activity. As the temperature increases, the bumping would get harder and, it is conceivable, some would bump hard enough to tear off electrons.

A thin conductor with sufficient current to reach the temperature necessary to "boil" off electrons is undergoing the thermo-electric effect.

If such a hot filament were placed in a vacuum near a strongly positive charged anode, the free electrons would be accelerated to the anode.

A simple demonstration can be arranged as follows (see Figure 16-8.):

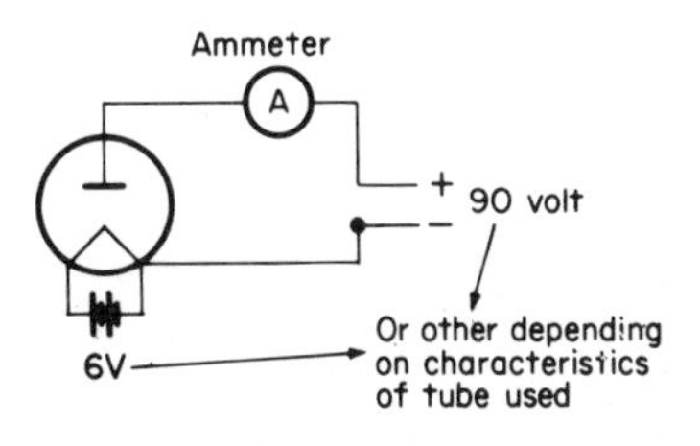

Figure 16-8

A diode tube set-up as in Figure 16-8 will permit a current to flow through the ammeter when the anode is positive relative to the cathode, but no current will flow when the polarity is reversed.

If you put variable resistors in the plate and filament circuits, you can also demonstrate the affect of varying voltage drops on the current in the plate circuit. (See Figure 16-9.)

A demonstration triode can also be assembled. (See Figure 16-10.)

This permits demonstrations similar to those performed with the diode but would also allow you to demonstrate the affect on the current in the plate circuit by a change in grid bias.

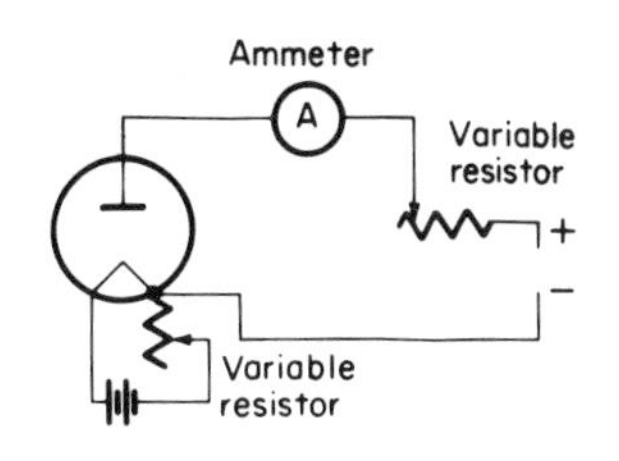

Figure 16-9

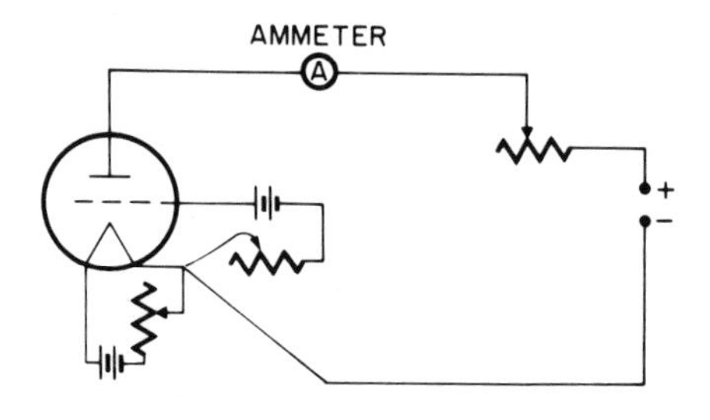

Figure 16-10

17

Magnetism

MAGNETISM IS A PHENOMENON most students find fascinating. Your students would probably have already sprinkled iron filings around bar magnets to illustrate lines of force. It might be well to review this demonstration and draw attention to the analogous configuration previously witnessed when demonstrating the electric force fields between charges with grass seeds floating on carbon tetrachloride around charged electrodes.

Magnetic Field and Electric Currents

The tendency to investigate magnetic effects as isolated phenomena should be strictly avoided. One should introduce the study of magnetism by some simple demonstrations showing that a moving magnetic field produces a current in a closed circuit. A current-bearing wire is surrounded by concentric magnetic field lines and a fluctuating or alternating current gives rise to a fluctuating magnetic field.

Some simple demonstrations can be performed as follows:

Make a coil of wire (about 10 turns of approximately 1.5 inches in diameter) and connect the two ends of the coil to a sensitive galvanometer. Introduce the pole of a bar magnet into the coil and observe the current reading on the galvanometer as the pole enters the coil. Let the magnet sit quietly in the coil and draw attention to the fact that current has gone to zero, emphasizing the need for a *moving* magnetic field to induce a current and then withdraw the magnet causing a current to flow in the coil in the opposite direction, as indicated by the galvanometer.

If your galvanometer is not sensitive enough, increase the number of coils or use a solenoid with a large number of windings. (See Figure 17-1.)

Moving the bar in and out of the coil at different rates affects the magnitude of the current, and varying the number of turns in the wire will also have a demonstrable effect on the current registered by the galvanometer.

These are also useful experiences to draw upon as work continues in this area.

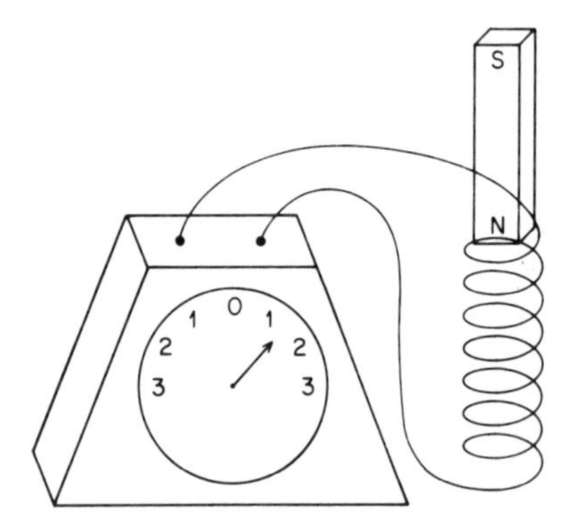

Figure 17-1

Set up a large coil (I have made a coil by running loops of wire around the stage of an overhead projector) and test the field around the wire, inside and outside of the coil, with a compass. When I demonstrate this with the overhead projector, I use a transparent compass which permits only the shadow of the compass needle to be projected on the screen. It then becomes very easy to introduce the right-hand rule, wherein if one uses the thumb of the right hand to point in the direction of the current flow (from positive to negative), the fingers bent at right angle to the palm point in the direction of the magnetic field, the direction in which the north-seeking pole of a compass points.

It now becomes incumbent upon the teacher to have his students investigate the phenomena just demonstrated in a laboratory situation in which they can find quantitative relationships between current in a wire or coil and the induced magnetic field. (See Figure 17-2.)

An apparatus built of wood with brass or aluminum fasteners and pegs, or made of some other non-ferrous materials, is satisfactory for this experiment. The apparatus designed for the PSSC experiments in this area is ideal for the purpose.

Students should be instructed to line up the apparatus so that when the compass is placed on the platform, the needle pointing in the direction of the earth's magnetic field is parallel to the plane of the coil around the platform.

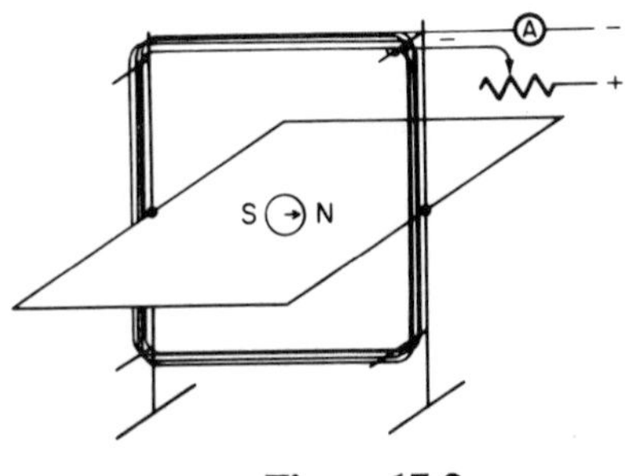

Figure 17-2

A current from a small voltage D.C. supply is then fed through the coil. Approximately 2–3 volts are satisfactory for this experiment. The current can be varied with the variable resistor in the line.

Since the effect on the compass needle is the result of two magnetic fields (the field of the earth which has been set parallel to the plane of the coil and the induced field of the coil which will be perpendicular to the plane of the coil), the forces on the compass needle can be added up vectorially. (See Figure 17-3.)

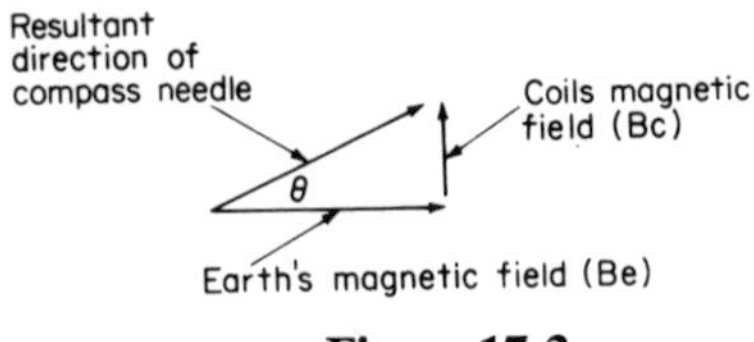

Figure 17-3

The coil's magnetic field can, therefore, be expressed as $Bc = Be \tan. \theta$, and by measuring the angular deviation of the compass needle from true north, the student can determine the strength of the induced field in units of earth magnetic fields.

Caution: Currents large enough to cause an approach to a 90° deviation in the direction of the compass needle from north are undesirable since the 90° deviation is approached as an asymptote with smaller and smaller increments of change as the current is increased. This would make the result of large currents very difficult to read and plot.

If the student then plots the current in the coil versus the tan of the angle of deviation, he should arrive at a linear relationship which should lead him to conclude that the induced magnetic field in a coil is directly proportional to the current flowing in the coil.

Placing polar graph paper on the platform with the compass in the center might help students read the angle with greater ease, but this is unnecessary if large compasses with plainly marked azimuth scales are used.

A similar experiment can now be performed by having the student maintain a constant amperage. Start off with approximately six coils of wire around the platform and plot the tan θ vs. the number of coils.

This should lead to the conclusion that the magnetic field in a solenoid is proportional to the number of coils in the solenoid.

After the experiment, students might be encouraged to "see" how predictable this result was. After all, if one coil produced a certain field, another identical coil with the same current in it should produce the same size field. If the two coils were lined up so that both fields point in the same direction, that would add up vectorially to give a resultant of twice the magnitude of the field produced by one coil.

Using essentially the same apparatus, students can now explore the field around long straight current-bearing conductors.

Unstring the loops from the compass platform and arrange the wire so that it is perpendicular to and alongside a table surface. (See Figure 17-4.)

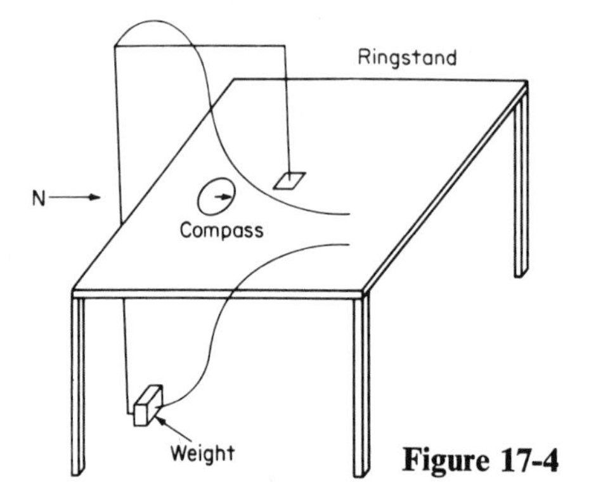

Figure 17-4

With a current flowing in the wire, students can explore the field around the wire with a compass, ascertaining for themselves that the magnetic field lines around a current bearing wire are concentric to the wire.

They can also verify the right-hand rule by pointing the thumb of the right hand in the direction of the current along the wire and seeing that the bent fingers of their right hand (the thumb is extended at right angles to the fingers and parallel to the plane of the palm) are in the direction indicated by the compass needle.

One can then proceed to measure the field as a function of distance from the wire by placing the compass on the table so that it points directly to or away from the wire, and then with a constant current flowing in the wire record the tan θ at various distances from the wire. Plotting tan θ vs. distance should indicate an inverse relationship between distance and the magnetic field strength around the wire.

The magnetic field ($\vec{B}$) induced by a current of 10 amp. in a loop 2π cm in radius is one gauss. If you have your students make stiff wire loops 2π cm in radius, they can use them to measure the strength of the magnetic field in your laboratory by placing a compass in the center of the loop and arranging the loop so that the plane of the loop is perpendicular to the earth's surface and parallel to the magnetic field of the earth. Current is then fed into the loop until the compass is deviated 45° from north. It is then that $\tan\theta$ equals one and the field of the coil is equal to the field of the earth. The amperage divided by ten will then equal the magnitude of the earth's magnetic field in gauss.

The experiments described lead to the conclusion that the field (B) inside a coil is related to the current (I) and the number of loops (N) as follows: $B = kIN$. Also the field around a current-bearing wire is related as $B = kI/r$, where r is the distance from the wire (or the radius of the loop).

We can then determine the value for k since

$$1 \text{ gauss} = k \times \frac{10 \text{ amp.}}{2\pi \times 10^{-2}} \text{ meters}$$

$$k = \frac{2\pi \times 10^{-2} \text{ meters} \times \text{gauss}}{10 \text{ amp.}} = 2\pi \times 10^{-3} \frac{\text{gauss-meters}}{\text{amp.}}$$

These are awkward units. The value of a gauss is 10^{-4} newtons/amp-meter.

$$\text{Therefore, } k = 2\pi \times 10^{-3} \frac{\text{meter}}{\text{amp.}} \times 10^{-4} \frac{\text{newtons}}{\text{amp.-meters}}$$

$$k = 2\pi \times 10^{-7} \frac{\text{newtons}}{\text{amp.}^2}$$

$$\text{and } B_e = 2\pi \times 10^{-7} \frac{\text{newtons}}{\text{amp.}^2} \times \frac{I \text{ amps}}{r \text{ meter}} = 2\pi \times 10^{-7} \frac{I}{r} \frac{\text{newtons}}{\text{amp-meter}}$$

The field inside a coil is-

$$B = 2\pi \times 10^{-7} \frac{I}{r} \frac{\text{newtons}}{\text{amp-meter}} \times n$$ (where n is the number of loops in the coil.

It is obvious that the field inside a loop r meters in radius is greater than the field r distance from a straight wire carrying the same current as the loop. This is because all sections of the loop are r meters from the point where the field is being measured, and thus the entire loop's current contributes equally to the field. In the case of the straight wire, only the perpendicular distance to the wire is r, the rest of the wire gets further and further from the point where the field B is being measured.

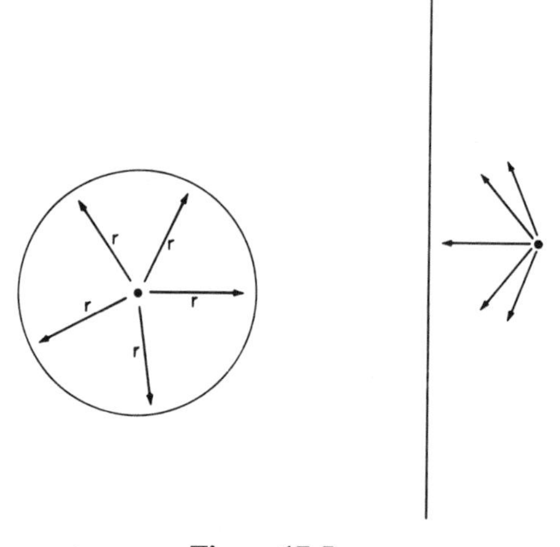

Figure 17-5

See Figure 17-5. An experiment designed to measure the field r distance from a current-bearing wire and compare it to the field in the center of a current-bearing loop (r) in radius (by techniques similar to those outlined above in the student experiments) would show that the field around the wire is $1/\pi \times$ the field inside the loop. Thus the field around the wire is

$$B = \left(2 \times 10^{-7} \times \frac{I}{r}\right) \frac{\text{newtons}}{\text{amp-meter}}$$

It is now time to examine the effect of a magnetic field on a current.

We have previously demonstrated that a moving magnetic field causes a current to flow in a conductor, and conversely we have demonstrated that a current causes a magnetic field to exist. Since the field engendered by the current is proportional to that current, a pulsating current would in turn create a pulsating magnetic field.

We can begin our lessons in this area by several interesting and simple demonstrations.

A very useful device is an unshielded cathode ray tube. (Most oscilloscopes are too well shielded to be affected by an external magnetic field.) An oscilloscope removed from its metal casing might work quite well for this demonstration.

Turn on the beam and point out that the spot on the oscilloscope screen is caused by the stream of electrons striking the fluorescing pigment painted on its inner surface.

Point out that when electrons are flowing from the "electron gun" in the rear of the tube to the screen in the front, this is the equivalent of positive current flowing from the screen in front of the tube to the electron source in the rear.

Bring the north pole of a magnet up toward the "cathode ray" tube (perpendicularly to the electron stream) and note the deflection of the stream.

Indicate the right-hand rule by showing that when the thumb of the right hand points in the direction of the current (opposite to direction of electron flow in tube) and the fingers point in the direction of the magnetic field (in this case in the direction of the moving north pole), then the palm faces the direction of the force exerted on the charges by the magnetic field.

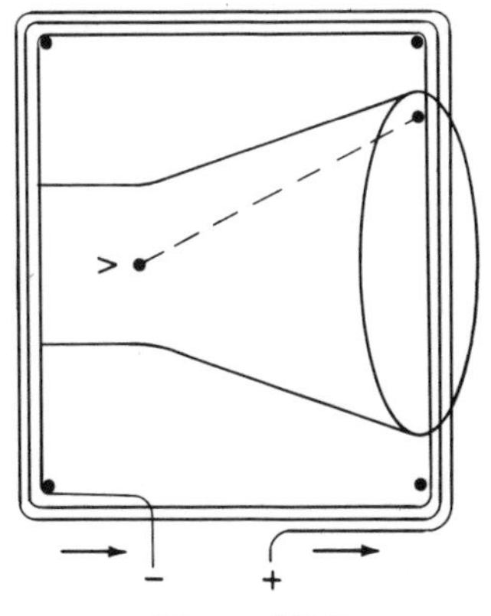

Figure 17-6

The same effect can be demonstrated by wrapping several coils of wire around the oscilloscope tube as in Figure 17-6. With current flowing counter-clockwise in the coils, as indicated, the magnetic field (B_1) for the loop will be pointing out of the page. With the fingers of the right hand pointing out of the page and the thumb pointing in the direction of the + current in the tube (toward the rear), the stream of electrons must be deflected up-wards as shown.

Field in Solenoid

A high priority student laboratory for devising the magnitude of the force relative to the currents flowing is the one described in the PSSC lab manual entitled "The Measurement of a Magnetic Field in Fundamental Units." The apparatus is arranged as illustrated in Figure 17-7.

The loop is inserted in the coil and balanced on the pivots. One ampere is fed into the loop and a variable current is fed into the coils in such a direction as to exert a down force on the section of the loop inside the coil. The force can be measured by hanging weights on the external end of the loop.

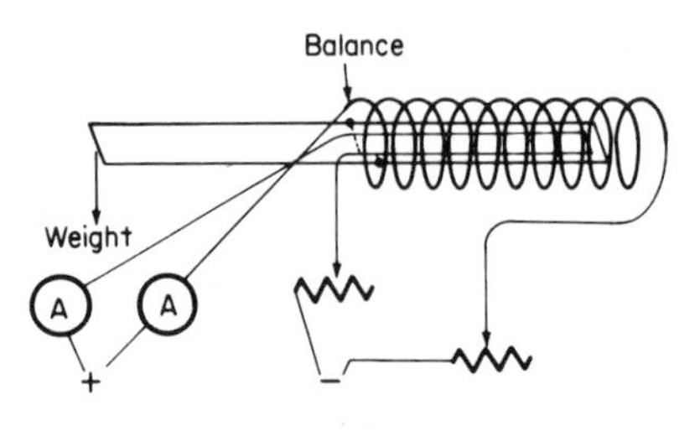

Figure 17-7

The field (B) in a coil has been found in terms of newtons/amp.-meter or $B = F/IL$ where F is the force acting on a current-bearing wire of length (L) carrying current (I). A solenoid with a constant number of coils will give rise to a field B directly proportional in magnitude to the current flowing in the coils.

This experiment is designed to measure the relationship of the current in the solenoid to the field generated.

The length of the portion of the inserted loop which is perpendicular to the axis of the solenoid is measured. This is the only portion of the loop affected by the field. The field generated by the solenoid is, of course, parallel to its axis.

The experiment can be varied by keeping the field of the solenoid constant and varying the current in the loop. Since $F = BIL$, it is evident that the force on the current-bearing loop should be proportional to that current.

Another useful variation, especially if you are short of ammeters, follows. Use the hook-up in Figure 17-8.

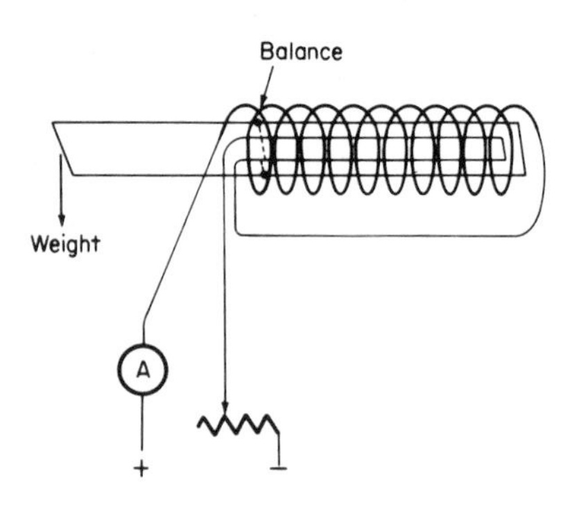

Figure 17-8

Connect the loop and the solenoid in series. The current in the loop and solenoid will now always be the same. Since $F = BI_eL$ (I_e being the current in the loop) and $B = kIN$ for the solenoid.

Therefore, $F = kI^2N$; that is, the force should be proportional to the square of the amperage if the loop and solenoid are in series.

Induction and Lenz's Law

An interesting display of inductance, which helps focus student attention, can be made with an easily constructed repulsion coil.

I made one by making the core from straightened pieces of wire coat hangers about eighteen inches long made into a bundle about one inch in diameter.

Around the bottom of this bundle, I wrapped 100 feet of bell-wire with a push-button switch and attached a regular two-prong electric A.C. plug. A ring of aluminum was then placed over the core, and when the coil was plugged in and activated by pressing the switch button, the aluminum ring was seen to rise and remain suspended above the coil as long as the A.C. current was on.

Students were then asked to volunteer an explanation for the observed effect.

The 60 cycle A.C. current in the coil causes, of course, a moving magnetic field perpendicular to the current (parallel to the axis of the coil). As shown previously, a magnetic field moving perpendicularly to a ring of wire will cause a current to flow in the wire (in this case the aluminum ring). The current produced in the ring, however, will give rise to a magnetic field which opposes the field in the coil. This effect is known as Lenz's Law. In effect, it is really predicted by the conservation of energy. To produce a current in the wire or the ring, work has to be done on it, and as the work is done, energy is lost in the loop by the agency of the opposing magnetic field produced in the ring.

Often we point out that when a wire is pushed through a magnetic field "cutting across the magnetic lines of forces," a potential is generated through the wire. The action can be much more easily visualized by applying the concepts illustrated throughout this section.

A wire may be considered to be an array of charged particles (the protons and electrons making up the metal atoms).

Let us investigate the action on the positive charges as we push the wire into a magnetic field remembering that the negative charges would move opposite to the direction of the current. (See Figure 17-9.)

Using the right-hand rule, as the wire is pushed to the right into a (B) field pointing downward, the thumb of the right hand would point to the right, the fingers would point downward, and the (E) electric field generated in the wire would be shown by the direction faced by the palm (into the page). Remember the field $E = F/q$ (force per unit charge) and

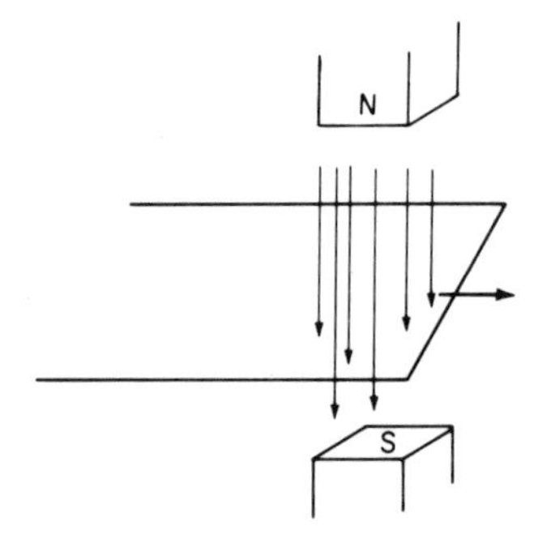

Figure 17-9: Electrons moving to the right are equivalent to currents to the left. Positive charges moving to the right are equivalent to negative charges moving to the left

$F = Bq\,v$, so $F/q = Bv$. The intensity of the field would be proportional to both the magnitude of the magnetic field and the velocity with which the wire is injected into the field.

If the two ends of the wire are connected by a conductor making a closed circuit, a current will flow in the direction of the field (the direction of positive flow—opposite to the direction of electron flow in a metal conductor).

If a current is flowing into the page, the thumb of your right hand now points in that direction, the fingers still point downward, and the palm faces now to the left. This shows that force on the positive charges in the wire is to the left opposing the motion moving them into the (B) field. This makes it necessary to do work to push them in, and the faster the flow of current, the greater the opposing force and the more work that must be done to maintain the flow.

The concept of an alternating current causing an alternately expanding and contracting magnetic field can be applied to such practical use as the transmitting and receiving of radio signals from one antenna to another.

The concept of electro-magnetic radiation in general ought now to be discussed.

For instance, an ion moving at constant velocity would radiate a stable magnetic field, another ion sitting in the stable magnetic field would feel no electrical field and thus would not be caused to move and generate a repulsive field to the original ion thus slowing it down. Only an accelerating ion would thus indirectly cause the repulsive field to be generated. We usually express this idea by saying that accelerating charges radiate away their energy.

If there were only one charge in the universe, it could accelerate without radiating away energy since there would be no other charges to pick up this energy. As the charge accelerates, it causes a changing magnetic field to radiate at right angles to its motion. When the fluctuating magnetic field crosses a charged particle in its path, an electric field is produced at right angles to the direction of the moving magnetic pulse. The resulting motion of the charged particle thus produces a magnetic pulse which, when crossing the path of the original ion, produces an electric field opposing its motion, thus slowing it down, taking away some of its energy. Thus stars radiate energy in the electro-magnetic spectrum as charged particles all through the universe accept this energy.

The vibrating ions in the plasma of stars colliding, moving back and forth in all possible directions with speeds and frequency of change of path and direction dependent on the temperature of the star, give rise to the entire spectrum of frequency of pulse generation we know as the electro-magnetic spectrum.

Circulation

We can experimentally verify that the field around a current-bearing straight wire is proportional to the current and inversely proportional to the distance from the wire, or $\vec{B}_{st}$ wire $= k\, I/r$. We have previously shown that the k in this relation is 2×10^{-7} newtons/amp.^{2}.

This field encircles the wire and at any point on the circular path around the wire is always tangential to that path. (See Figure 17-10.)

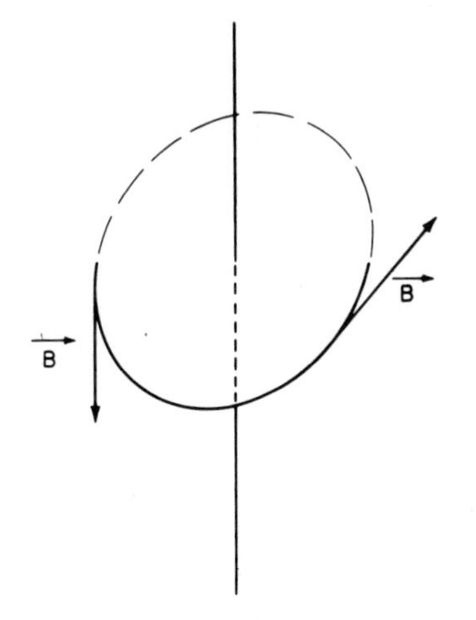

Figure 17-10

The length of the path of this field intensity times $\vec{B}$ is defined as the circulation of $\vec{B}$,

and if $\vec{B} = k\,\dfrac{I}{r}$

$$\vec{B}r = kI$$

and $\quad \vec{B} \cdot 2\pi r = 2\pi kI$

Since $\vec{B}$ is inversely proportional to r, it can be seen from the above that the circulation has the same value at any distance (r) from the wire as long as the current (I) is constant. The circulation varies proportionally to that current.

$2\pi k$ is sometimes called μ_0, so that we can write $\vec{B}(2\pi r) = \mu_0 I$. This expression can be generalized and shown to be applicable also to more complicated situations. It is called Ampere's Law.

We may restate it by saying that $\vec{B}(2\pi r) = \mu_0 I$ where I is a current threading through a surface bounded by $2\pi r$.

Thus, as long as a current (I) goes through this surface bounded by $2\pi r$, the circulation ($B \cdot 2\pi r$) will have some value, depending on that current (I). It is only when no current threads through this surface that the circulation becomes zero.

Furthermore, the shape of this surface need not be round as long as it is bounded by a perimeter equal to $2\pi r$. (See Figure 17-11.)

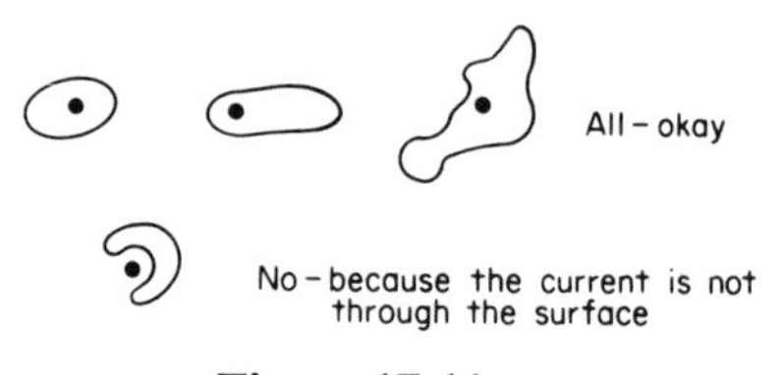

Figure 17-11

The argument for the independence of the shape of the surface can be phrased as follows (Figure 17-12):

$$\vec{B} \cdot 2\pi r = B \cdot \Delta d_1 + B \cdot \Delta d_2 + B\Delta d_3, \text{ etc.}$$

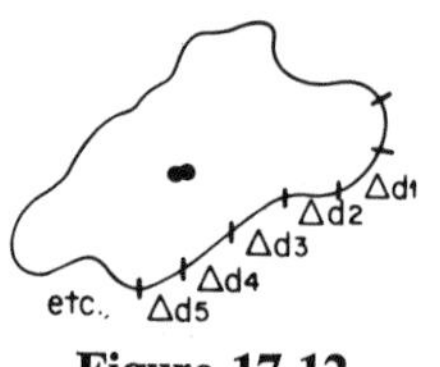

Figure 17-12

Since B varies inversely with r (the distance from the current-bearing conductor), then when the Δd being measured is far from the conductor, the $B\Delta d$ will be small, and when the Δd being measured is close to the conductor, the $B\Delta d$ will be large. They will all average out so that the sum of all the $B\Delta d$'s will be equal to $B \cdot 2\pi r$ as if all the Δd's were an equidistance, r, from the conductor.

However, when the conductor is outside, the enclosed surface is as follows (Figure 17-13):

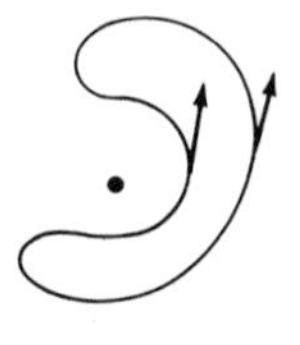

Figure 17-13

When you add up the $B\Delta d$'s either clockwise or counterclockwise, some of the $\vec{B}$'s (all tangential to the perimeter) will be pointing in the clockwise direction and some will be counterclockwise so that there will occur a resultant cancellation and the sum of the $B\Delta d$'s will be zero.

Field Inside a Solenoid (See Figure 17-14.)

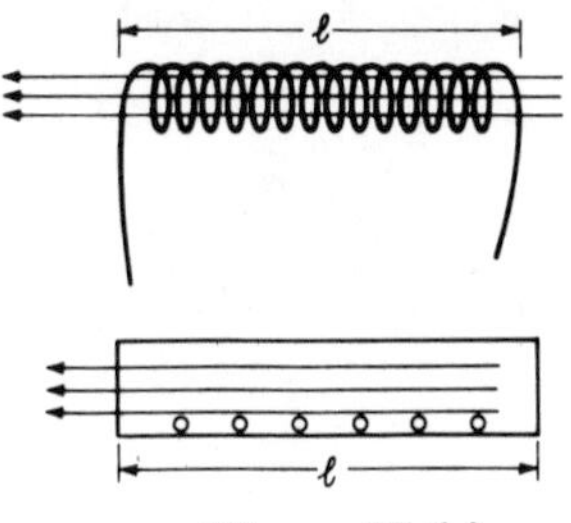

Figure 17-14

Note that inside a solenoid the field is uniform. Field lines are parallel (analogous to the situation wherein the electric field between the plates of a capacitor is uniform). The field outside the infinitely long solenoid is zero.

The circulation for the rectangular area parallel to the axis of the solenoid and threaded by n current-carrying wires is, therefore,

$$Bl = 2\pi kIn$$

$$\therefore B \text{ inside solenoid} = 2\pi kI\frac{n}{1}$$

if $\dfrac{n}{1} = N$, then $B = 2\pi kIN$

It has previously been shown that

$$B_{st} \text{ wire} = 2 \times 10^{-7}\frac{I}{d} = k\frac{I}{d}$$

and $k = 2 \times 10^{-7} \frac{\text{newtons}}{\text{amps.}^2}$

The field inside a solenoid, therefore, is

$$B = 2\pi \cdot 2 \times 10^{-7} \frac{\text{newton}}{\text{amps.}^2} \times \frac{n}{\text{meter}} \times I \text{ amp.}$$

$$B = 4\pi \times 10^{-7} \frac{\text{newton}}{\text{amp. meter}} \cdot I \cdot n$$

But the field in the center of a loop of wire has been shown to be $B = 2\pi \times 10^{-7} I/r \cdot n$. Why?

Remember that the 2×10^{-7} for the straight line conductor is smaller than the $2\pi \times 10^{-7}$ for the loop because all of the loop contributes to the field at r distance away, while a small portion of the straight line makes a strong contribution. In the case of the solenoid of infinite length, the field with the factor $4\pi \times 10^{-7}$ is twice that of the $2\pi \times 10^{-7}$ factor for the loop because many coils contribute to the field with no end effects. In the case of the loop, the ends where there is no continuity of coils make no contribution.

18

Practical Electricity

THE PRACTICAL ASPECTS of electricity may now be examined in detail if you so desire. Transformers, tuners, motors, meters, and generators are all usually discussed in elementary physics courses. I do not propose to go into much detail here. I will merely point out the applications which might help make the study of basic concepts more interesting.

We had previously shown that the heat generated in a current-bearing wire changes as the square of the amperage (cal $= .24\, I^2Rt$) and only directly as the voltage (cal $= .24\, EIt$). It is, therefore, advantageous, if we are going to transmit electrical energy for long distances, to send it out at high voltage and low amperage and then raise the amperage and lower the voltage as required by the user. We can do this with alternating current and suitable transformers. Direct currents are, as we can now clearly point out, unsuitable for such manipulations.

The use of transformers in radio where we need various voltages for filament heating, plate potentials, etc. can also now be easily understood. It is basic that the same concepts involving electro-magnetic radiation in general are applicable to radio signal transmission and receiving, transformers, and numerous other specialized aspects that you and your students will probably think of.

The principle of self-inductance and back voltage might be explained at this point.

When alternating or pulsating current flows through a solenoid made of adjoining coils connected in series, each coil of the solenoid is subjected to the magnetic flux generated by the currents in its neighbors. According to

Lenz's Law, this will induce e.m.f. which will act as an impedance to the flow of current in the solenoid. The greater the frequency of the A.C. current, the higher the resulting impedance. A solenoid or choke coil then becomes a natural for filtering high frequency A.C. current out of a circuit. D.C. current and very low frequency A.C. can pass through such a coil easily.

A capacitor, on the other hand, is very useful for permitting high frequencies but will prevent the passage of low frequency A.C. This is easy to explain by pointing out that as the current continues to flow in one direction, the charge on the capacitor continues to build and the resulting capacitative reactance tends to produce a back e.m.f. which prevents the current from continuing to flow. Thus for lower frequencies, large capacitors may permit current to flow since reactance would build up slowly, while in a small capacitor the reactance would build up more quickly than the time required for the A.C. to reverse itself. The smaller the capacitor, the higher the frequency that is therefore required for continuous current flow.

Variable capacitors, as in tuning circuits, permit us to select the frequency we wish to flow through our radio circuits.

Generators

We have already illustrated the generation of an A.C. current by pushing a bar magnet in and out of a coil of wire.

Let us now investigate the results of rotating a coil of wire in a magnetic field. (See Figure 18-1.)

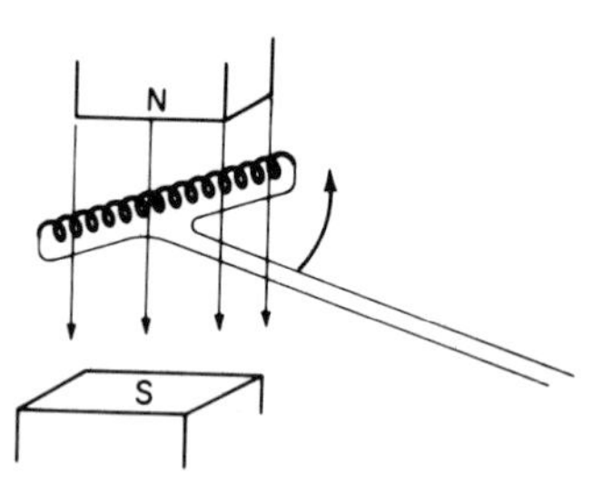

Figure 18-1

As the coil rotates, the component of its motion varies from being parallel to the magnetic field to being perpendicular to the field. Remember the situation in our derivation of simple harmonic motion. (See Figure 11-1.)

If we plot the perpendicular component of velocity against time, we get a graph which we should recognize as a sine curve. We have actually been plotting the sine of the angle of rotation as the coil rotated through

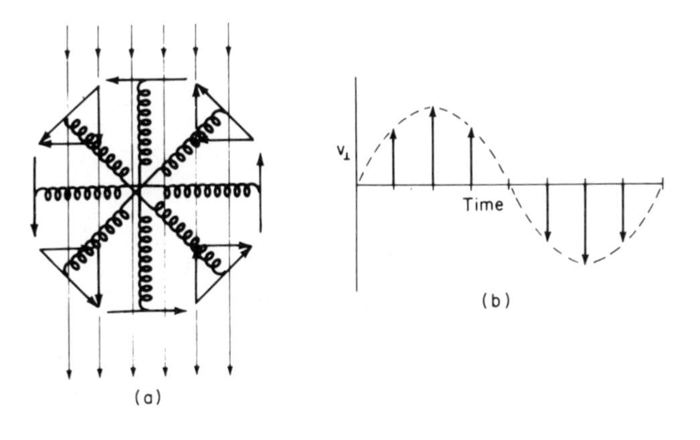

Figure 18-2

one complete rotation. (See Figure 18-2.) Since emf $= B_{\perp} vl$, the e.m.f. generated across the coil is directly proportional to this perpendicular component of velocity and will, therefore, vary in the same manner.

But $E = IR$, and if the output circuit of this generator is closed through a resistance r, the current will vary directly with the e.m.f. and an A.C. current will result.

If an A.C. current is desirable, the current to the external circuit is taken by contacts made with slip rings on the shaft of the generator rotor. (See Figure 18-3.)

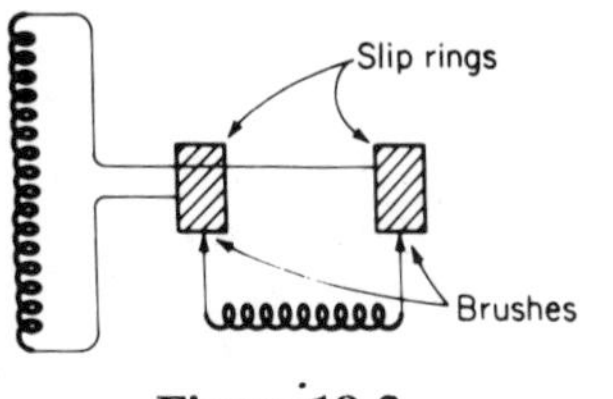

Figure 18-3

If, on the other hand, a D.C. current in the external circuit is desired, the brushes of the external circuit are held in contact with a split ring called a commutator. (See Figure 18-4.)

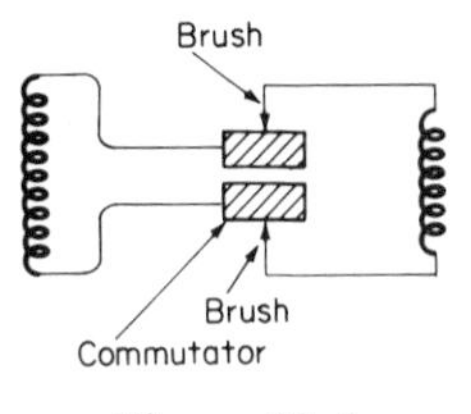

Figure 18-4

Now the commutator rotates with the rotor and when the current reverses as the rotor moves through the field, the commutator half is now contacting the other brush so that the current in the external circuit is always D.C., albeit pulsating D.C. (See Figure 18-5.)

Figure 18-5

To smooth out the pulsations, one may have several rotors going through the field with the commutator split several times so that the current does not have time to fall to zero before the brushes are in contact with another section of the commutator. One can also fill in the valleys and smooth out the peaks on the curve by installing a high resistance and a grounding capacitor on the downstream side of the generator. (See Figure 18-6.)

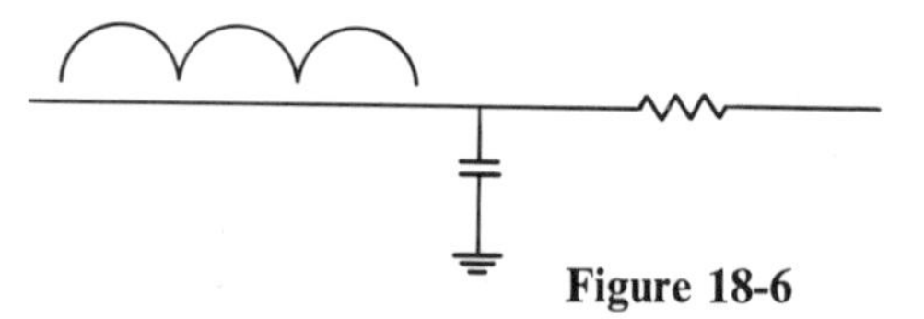

Figure 18-6

Thus when the current is high, the resistor will cause the capacitor to become loaded, and when the current is low, the capacitor will unload its excess charge into the mainstream, filling out the low spots in the current curve.

Motors

A motor is, of course, a generator in which current is fed into a coil causing it to rotate in a magnetic field, thus converting electrical energy into mechanical energy. Using essentially the same figures used to illustrate the generator in the previous paragraph, one can show that if an A.C. current is fed into the slip rings of the rotor through the brushes of the external circuit, the coil will become an electromagnet with one end attracted to one side of the field magnet. (See Figure 18-7.) When it is rotated by this attraction, its inertia will carry it a little past and the A.C. current will then reverse itself so that instead of the coil stopping and turning back, it will continue revolving. If a D.C. current is used to power this motor, then the current is fed in through a commutator and the reasoning is the same as in the description of the commutator in the generator.

The fact that a motor and generator are the same thing with the flow reversed can now be used to show why a motor heats up and may burn out when it is overloaded.

To start the motor up, current is fed into the rotor. As the rotor begins to revolve in the field, a back e.m.f. begins to be induced in the rotor coil

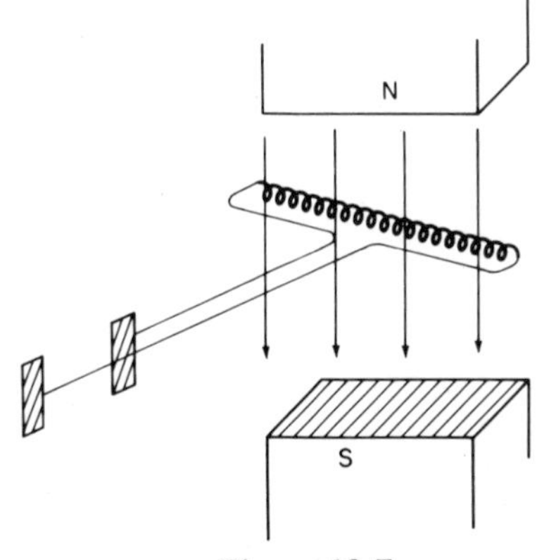

Figure 18-7

(remember Lenz's Law). The faster the coil rotates through the field, the greater becomes this self-inductance which, when added to the natural resistance of the coil wires, becomes the total impedance retarding the flow of current. It therefore is easily seen that the faster the coil rotates through the field, the less current will flow through its coils. Of course an equilibrium is soon reached because, as it begins to slow down, more current will flow causing it to speed up again. It now becomes evident why lights sometimes dim momentarily when a motor is started up, since we see it takes more current to start the motor up than to keep it going, thus leading to a momentary drop in voltage.

What happens when a load is put on the motor which prevents it from turning as fast as it should to build up the necessary inductive reactance to keep the current low? Since heat generated in the coil varies as the square of the current (cal $= .24\ I^2rt$), the coil will very quickly heat up and, unless protected by a fuse or circuit breaker, will burn up.

Radio Diode

Some further applications to radio technology might be interesting to some of your students. The diode or rectifying vacuum tube is easy to start with. It might be interesting to note that the diode vacuum tube was originally developed by accident in the Edison Laboratories.

In the early incandescent bulbs, one of the problems encountered in the evacuated bulb was the evaporation of filament metal by the heat and the subsequent deposition of the metal vapor on the relatively cool glass surface, causing the bulb to darken as the glass became opaque. To solve this problem, a heavy metal plate was placed inside the bulb with a metal conductor through the glass. (See Figure 18-8.)

It was hoped that the superior heat conducting properties of metal would help keep the interior of the bulb a little cooler and increase the use-

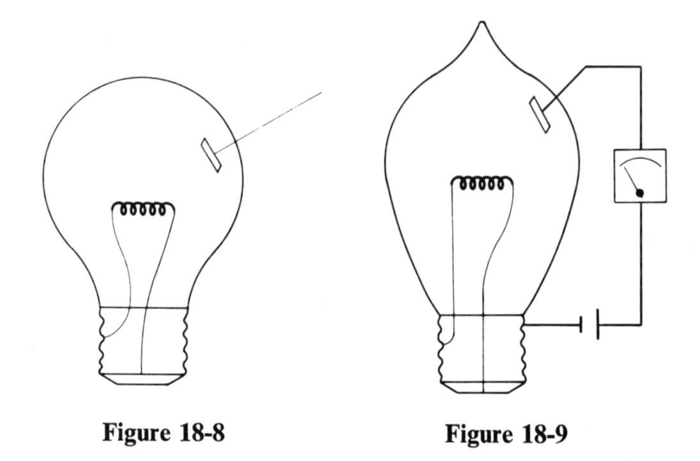

Figure 18-8 Figure 18-9

ful life of the bulbs. It did not prove to be practical. However, Edison and his co-workers proceeded to do some experiments with this new apparatus. It was wired as in Figure 18-9.

An A.C. current was fed into the filament to keep it incandescent. The D.C. source was connected to the plate and filament through an ammeter. It was noted that when the filament was connected to the negative side of the D.C. source and the plate was positive, a current flowed in the ammeter. However, when the D.C. source was reversed so that the filament was positive with respect to the plate, no current was registered on the ammeter. You might ask your students to explain this.

The atoms in the hot filaments are undergoing intense thermal motions and violent collisions are taking place, violent enough to tear off electrons. This boiling off of electrons from a hot filament is known as the thermo-ionic effect. When the plate is positive, electrons are attracted to it and a current is registered in the ammeter. When the plate is negative, electrons are, of course, repelled and no current is found to flow through the ammeter.

This was duly reported in the journals of the day and was called the "Edison Effect." During this period, radio was in its infancy.

Radio signals were transmitted by superimposing a sound pattern on a constant frequency carrier wave so that an amplitude modulated wave arrived at the radio antenna (most commercial broadcasting is still via A.M. transmission).

An alternating current is, therefore, induced in the tuning circuit of the receiver. (See Figure 18-10.) Such a signal, with carrier waves frequency in the megacycle range and sound frequencies supered on the top and bottom of the wave cycle, is not reproducible by diaphragms of earphones. The

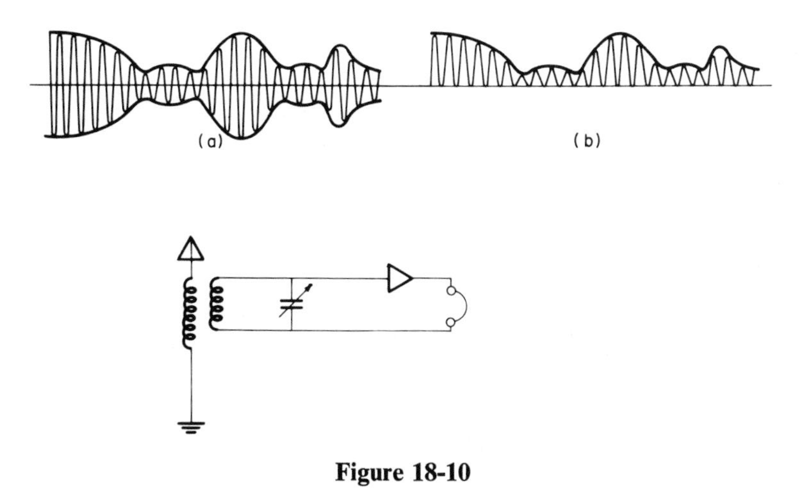

Figure 18-10

diaphragm cannot, of course, vibrate in the megacycle range, and as for the supered audio frequency, it is, in effect, being pulled in opposite directions at the same time. To make it possible for the earphone diaphragm to respond, the signal must be rectified; that is, the current must be caused to flow in one direction only. This was done with the use of lead sulphide "Galena" crystals which have the property of permitting current to flow in one direction only, similar to the way a check valve works in a water line.

Some of your students may have played with such crystal sets. Perhaps they remember hunting for a sensitive spot on the crystal with the "cat's whisker." Next time the same spot was no good because oil from their fingers covered it so that they had to hunt for another spot in order to get a signal.

Fleming saw the practical aspects of the "Edison Effect."

The tube is placed in the circuit (see Figure 18-11) in place of the galena

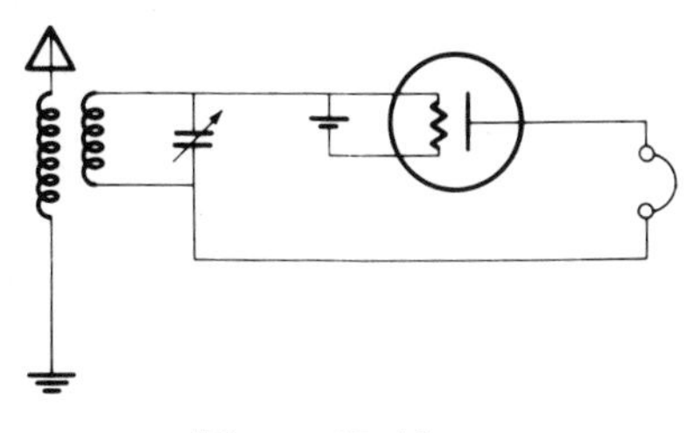

Figure 18-11

crystal. Current only flows through the earphones when the plate is positive though an A.C. current flows in the tuning circuit. The signal is now rectified.

(See description of diode and triode demonstrations at the end of Chapter 12.)

A much more dependable receiver than the old crystal set was now possible.

Triode

A difficulty still existed. The strength of the received signal still depended on the set's proximity to the transmitter.

It was not until De Forest invented the triode that it became possible to amplify the signal and thus permit receipt of signals at long distance from the transmitter. (See Figure 18-12.)

The filament is heated by a small, approximately six volt, "A" battery as in the diode. A large e.m.f. (approximately 100 volts) is then placed between the plate and the filament via a "B" battery. A steady strong current now flows through the earphones. A weak signal is fed from the tuning circuit to the grid of the triode. The grid is kept negatively biased with the "C" battery. Since the grid is much closer to the filament than the plate is, a very weak charge on the filament has a much greater effect on electrons leaving the filament than the strong charge the plate has on it because of the inverse square law. When the grid is negative no electrons flow to the plate, and when the grid becomes more and more positive, a larger flow of electrons to the plate becomes possible.

Thus a very weak charge on the control grid patterns the strong current in the plate circuit much like a seamstress using a tissue paper pattern to cut a shape out of a piece of cloth. The invention of the triode tube was the impetus which permitted radio to assume the important niche it has in modern technology.

SOLID STATE

Diode

Solid state technology became possible after research in the physics of the solid state clarified the properties of crystals. It is beyond the scope of this book to go into much detail, but some of the theory and practical applications might be summarized as follows:

A pure crystal of germanium or silicon is a non-conductor. If certain impurities are present, some conductivity is evidenced.

Germanium has four electrons in its outer shell. We can grow a germanium crystal from pure germanium to which is added a smaller percentage of gallium. Gallium has three outer electrons, and a representation of the arrangement of atoms in a germanium crystal doped with gallium might look something like Figure 18-13.

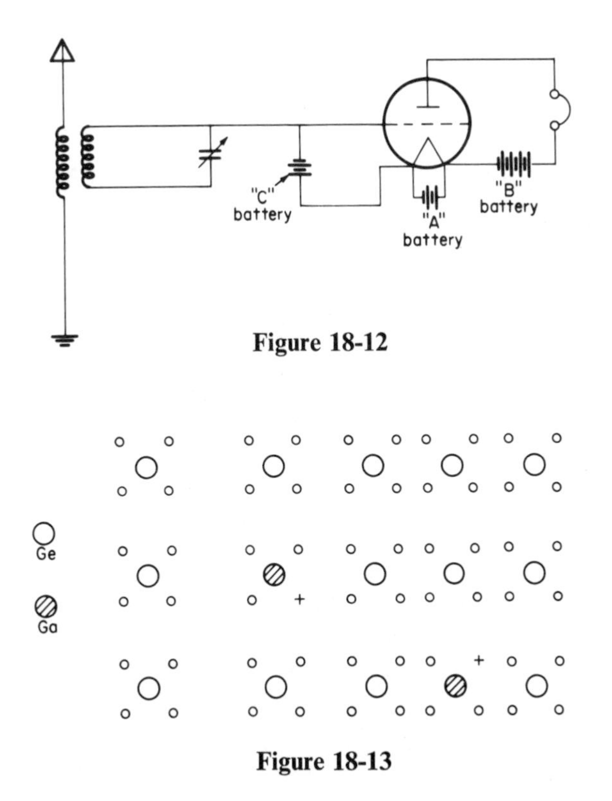

Figure 18-12

Figure 18-13

Note that each germanium atom is surrounded by four electrons, while the two gallium atoms in the Figure are each surrounded by three electrons. Now suppose we add electrical stress to the crystal by introducing a potential drop across it. That is, we attach it to a battery so that one side of the crystal becomes strongly positive while its other side becomes negative. If the crystal is pure germanium, nothing would happen, but with the presence of a gallium, electrons can move. There is room for this move because of the missing electrons in the lattice surrounding the gallium atom. When an electron from a neighboring germanium atom is pushed into this vacancy, the hole left behind is now the equivalent of a positive charge. Thus, as electrons move to the left under influence of the applied potential, holes (as positive charges) are seen moving to the right.

A crystal of germanium can be produced in which the added impurity is arsenic. Arsenic has five outer electrons, and the arrangements of atoms in such a crystal can be represented by Figure 18-14.

When a potential is introduced across such a crystal, the fifth electron on the arsenic atom would be pushed over onto a neighboring germanium atom on its left. This leaves a positive hole on the arsenic atom which might then be filled by an electron contributed by the germanium atom on its right, leaving the germanium atom with a positive hole. Thus again, we have negative electrons moving in one direction and positive holes moving in the other direction.

The next step is to grow a crystal of germanium, half of which is doped with a minute amount of gallium impurity (known as a p-type crystal because it will have an excess of positive holes when subjected to a potential drop). The other half will be n-type, in which the doping impurity is arsenic leading to excess electrons. (See Figure 18-15.)

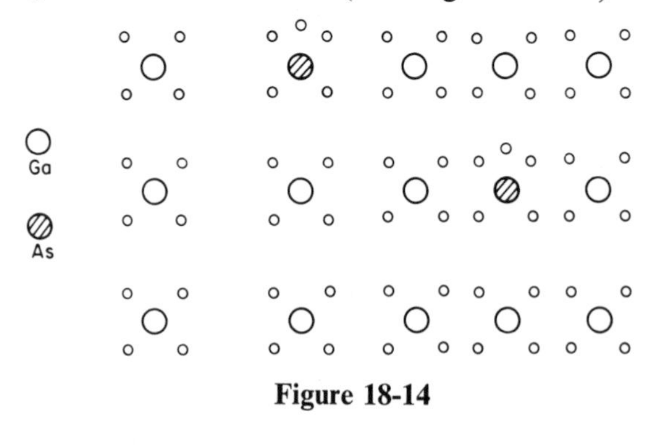

Figure 18-14

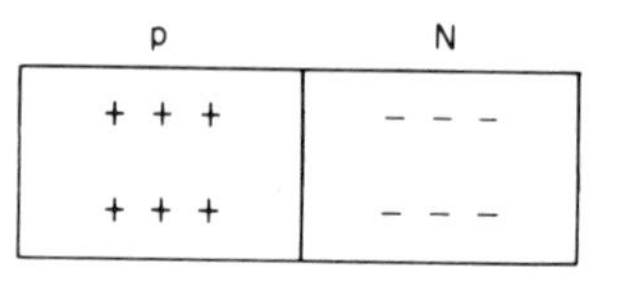

Figure 18-15

When a potential is placed across such a P-N junction, a current will flow only when the e.m.f. is in the right direction. This would be easy for your students to visualize when you suggest diagrams such as Figures 18-16 and 18-17.

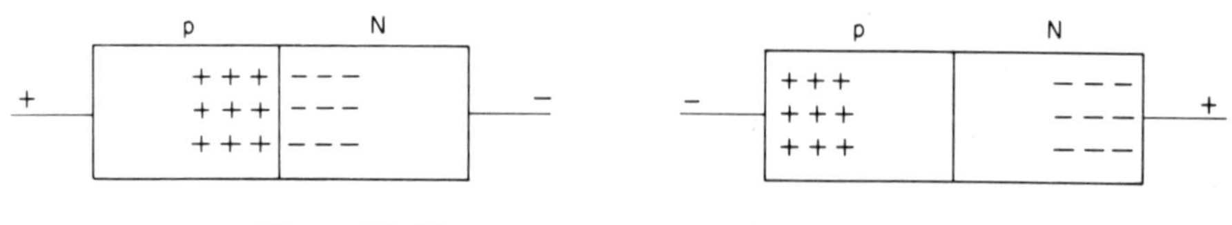

Figure 18-16 Figure 18-17

If the p-type side is made positive and the n-type is made negative (Figure 18-16), both the positive holes and negative excess electrons will accumulate at the junction. An exchange can take place and a current will flow.

If, on the other hand, the p-type side of the junction is made negative and the n-type side made positive (Figure 18-17), the excess electrons and positive holes would gravitate away from the junction. No interchange is then possible and no current will flow.

This crystal diode has the advantage of not needing a hot filament to boil off electrons. This cuts down on the energy required to operate the diode. It is also possible to make a very small crystal diode which will duplicate the work of a much larger vacuum tube, thus leading to effective miniaturization.

One disadvantage of crystal diodes is their temperature sensitivity. A vacuum tube with its heated filament has a self-contained temperature regulator, but a crystal diode must depend on atmospheric temperature or some external heater for its efficient operation.

Triode

The transistor is actually made up of two such junctions put together.

Consider a crystalline bar consisting of a p-type section, a very thin n-type section, and a p-type section. (See Figure 18-18.)

Assume the hook-up illustrated in Figure 18-19.

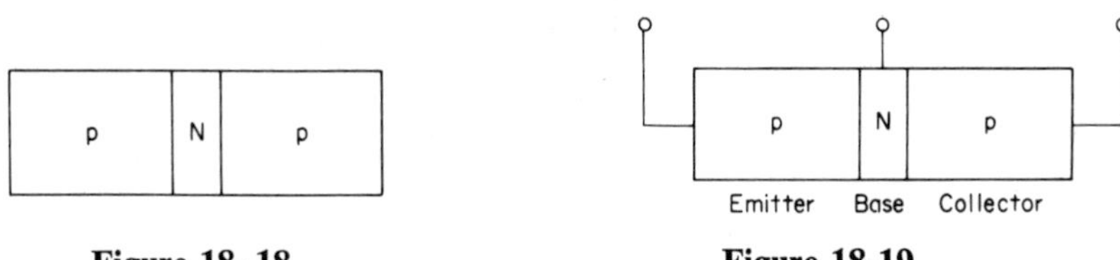

Figure 18-18 **Figure 18-19**

The first p-type section is called the emitter, the n-type section is called the base, and the second p-type section is designated the collector.

Potentials are then created via an outside source so that the base is negative with respect to the emitter and the collector is still more negative with respect to the base.

Since the base is negative with respect to the emitter, positive holes will flow from the emitter to the base. The base is made very thin, so as positive charges flow from the emitter into the base, some of these charges will diffuse across the base-collector junction and flow down the greater potential hill provided in the collector.

By varying the negative bias on the base (impressing the signal on it, much the same way the signal varies the bias of the grid in the triode vacuum tube), the number of positive holes entering the base from the emitter and diffusing across to the collector will be varied. Here again a very small change in the bias impressed on the base will cause a relatively large change in the current flowing into the collector.

The transistor thus serves as an amplifier. Transistors can also be made of N-P-N junctions in which the electrons carry the signal and all the potentials are reversed.

The above is, of course, a very simple explanation but will serve to satisfy the curiosity of your students and may, hopefully, induce some to dig further into the subject for themselves.

Mass Spectography

The relationships derived earlier and the knowledge that an elementary charge is equal to 1.6×10^{-19} coulombs can now be used for an experiment to discover the mass of an electron, or indeed to discover the mass of any charged particle.

Since $F = BIl$ and $I = q/t$, then $F = B\,q/t\,l$. But $l/t = v$, so $F = Bqv$.

The force on a moving charge moving through a field B perpendicularly to the direction of the field would then depend on the intensity of the field, the size of the charge and the velocity of the charge moving through the field.

The velocity of a charge of given mass is a function of its momentum.

Our experimental apparatus would require an electron gun with which we can impart a known amount of energy to the electrons and a magnetic field around which the electrons would be forced to circle. (See Figure 18-20.)

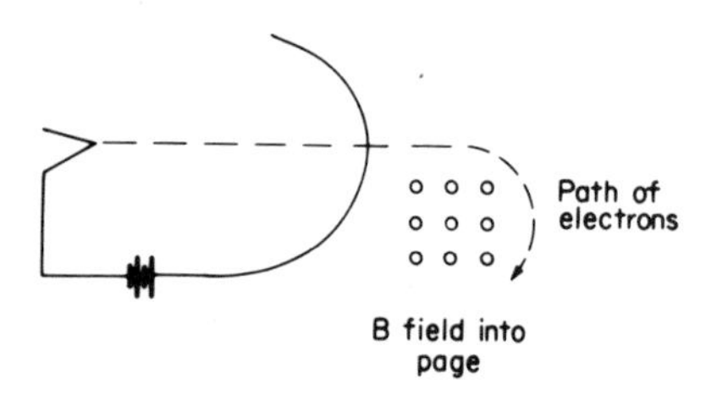

Figure 18-20

Since the force on the electrons (Bqv) causes them to move in a circle, it is a centripetal force and $Bqv = mv^2/r$ or $Bq = mv/r$ and mv (the momentum of the electrons) equals Bqr where r is the measurable radius of curvature of the electron path as it is forced around the magnetic field (B).

The energy with which the electrons enter the field is known since this is the energy imparted by the electron gun,

where

$$Vq = \tfrac{1}{2}\,mv^2$$

since

$$\frac{mv^2}{2} \times \frac{m}{m} = \frac{m^2v^2}{2m}$$

$$Vq = \frac{m^2v^2}{2m} \text{ and since } mv = Bqr$$

then

$$Vq = \frac{B^2q^2r^2}{2m} \text{ and}$$

$$m = \frac{B^2qr^2}{2V}.$$

Thus the mass of the electron can be determined in terms of known and measurable parameters.

This experiment is performed with very simple and inexpensive equipment in a PSSC laboratory exercise using the solenoid previously calibrated and an electronic vacuum tube.

Sometimes it is advantageous to perform this exercise as a class experiment using a large coil and a vacuum tube containing a small quantity of gas which will fluoresce showing the path of electrons.

This apparatus is available from laboratory supply houses. The coil is called a Helmholtz coil and is designed to produce an even measurable B field along its axis. The field (B) can be calculated by knowing the current, the number of coils, and the radius of the coil ($B = 2\pi \times 10^{-7}\, I/r$ newtons/amp-meter), or the coil characteristics can be supplied in a calibration curve giving the field B per ampere of current in the coil.

The energy supplied the emitted electrons can be calculated by knowing the potential from the cathode to the anode in the electron tube. A convenient device is usually supplied to make it easy to measure the radius of curvature of the electron path.

With the apparatus and instruments in full view of the class and a previously given lesson in which your students have become fully cognizant of the function of each component of your apparatus, you can now have them read and record the various parameters as you vary the voltage across the gun or the current in the coil. Knowing the size of the electron charge (as previously determined in the Millikan Experiment), you are now able to experimentally verify the mass of the electron.

The concept of mass spectrography can now be discussed in detail. For instance, how could you use these principles in a practical device to measure the ratio of U^{235} to U^{238} in a uranium sample or even to separate U^{235} from U^{238} in order to obtain a pure sample of the fissionable U^{235}?

A hot sample of U ions is permitted to leave an oven. A portion of this beam of ions is selected for equal velocity with a velocity selector made up of a capacitor and a coil. (See Figure 18-21.)

If the electric field E in the capacitor is upward and the magnetic field

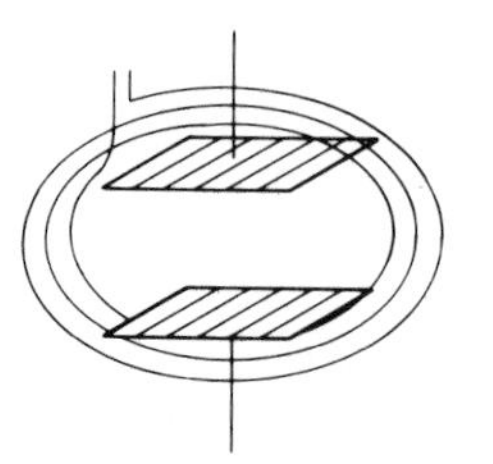

Figure 18-21

B due to the coil is downward, then when $Eq = Bqv$ only those ions moving through this field with a velocity $v = E/B$ will move through the field undeviated and will then enter the mass spectrograph where the radius of curvature of its path will be determined by its mass.

Since $mv = Bqr$, then $r = mv/Bq$, and since v, B and q are constant, the radius of the ion path would be proportional to its mass.

19

Photons and Matter Waves

MANY TEACHERS who do not finish the PSSC textbook have often protested that it is much better to teach in depth even if you do not finish the course. This is probably true, but it is still better to teach in depth and also finish the course if it is possible.

The last few chapters on modern physics describe some of the most exciting aspects of physics, and my students generally agree with me that all the pushing, cajoling and hard work was worth the effort. These last several chapters are like the dessert after a long heavy meal.

Introducing Quanta

Some of the chemistry teachers in our Science Department must teach quanta to juniors who have not had the background my senior physics students have had. I prepare the following as a hand-out for distribution to these students:

LIGHT IN QUANTA

Experiments during the nineteenth century seem to indicate that light apparently behaves as if it were a wave. We can even measure the wave length of light and can show that red light is longer in wave length than blue light.

We also know that the wave length of light times its frequency is equal to the velocity of propagation of the wave ($\lambda\nu = c$). (Velocity (c) of propagation—The velocity with which a wave moves through its media. Frequency (ν)—The number of waves that go by a given point per unit of time. Wave length (λ)—The distance between a point on a wave to a corresponding point on a succeeding or preceding wave.)

This is very reasonable since we can visualize that if 10 meter sticks pass by a given point every second, then the front end of the first meter stick must have been traveling at 10 meters per second in order to leave room for the 10 meter sticks following.

Now let us see how the wave nature of light would predict how light would react with matter, as in the photo-electric effect.

We can set up a piece of apparatus as in Figure 19-1.

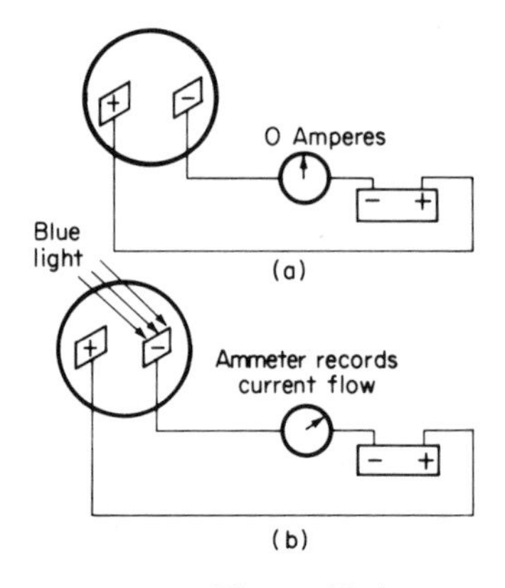

Figure 19-1

The two pieces of metal labeled + and − are enclosed in a vacuum and no current flows as shown in the ammeter (a device used for measuring an electric current). Now we bathe the (−) plate with blue light and we find a small current is flowing, as measured by the ammeter. We reason that the electrons in the metal electrode (−) have absorbed enough energy to leave the surface of the metal and be attracted to the positive plate and hence flow through the ammeter to the positive terminal of the battery.

As we increase and decrease the intensity (the amount of light per unit area on which the light falls) of the blue light shining on the negative plate, we find the current increases and decreases. We reason that more or fewer electrons are emitted as we add more or less energy in the form of light. So far, our wave model for light would seem to be satisfactory. Now we begin to cut the intensity of light bathing the negative electrode to lower and lower levels, and we observe something strange. The needle on the ammeter (and we must assume an extremely sensitive ammeter) begins to fluctuate. This might be recording the passage of single electrons. If light poured into the plate in a smooth continuum, as in a wave model, one would expect the pulses to be evenly divided, since it would appear that energy must be accumulated in order to release an electron. The interval of time between each electron release should be even since it should take just as long to accumulate the necessary energy to release each succeeding electron. This, in fact, does not happen. The needle fluctuates sporadically, and if we suddenly turned on the light, there is sometimes an immediate response. This does not sound reasonable—not enough energy could be accumulated if it were being added in a smooth continuum to justify an immediate response and then long and sporadic intervals between succeeding responses. It would appear more as though the light were arriving in little packets at random intervals; when the intensity of light increased, the intervals between arrivals of packets decreased, so that they appeared to arrive in a smooth continuum.

We continue our experiment using the same apparatus, modified slightly, as in Figure 19-2.

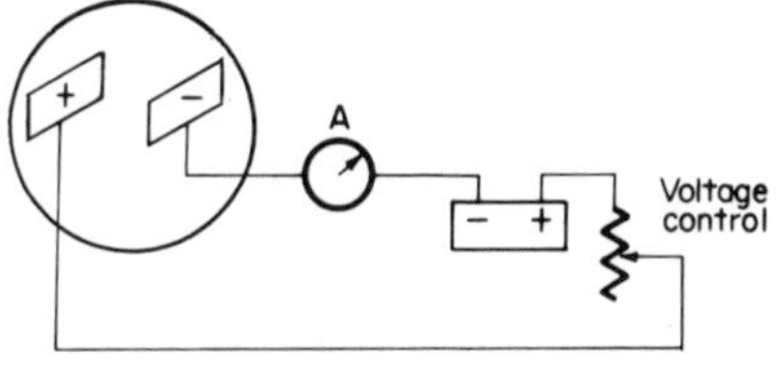

Figure 19-2

We now can vary the voltage across the electrodes. Blue light is directed now against the *positive* surface, and the voltage is increased until the current stops. (The electrons have to be projected to the negative plate and may be thought of as going uphill; therefore, they must have enough energy to climb up this potential hill.) When the current is off, a light of shorter wave length (maybe ultra-violet) is shined on the surface. Again, the current starts; again, the voltage is increased until the current stops. This is continued and the data obtained is graphed. The graph obtained is illustrated in Figure 19-3.

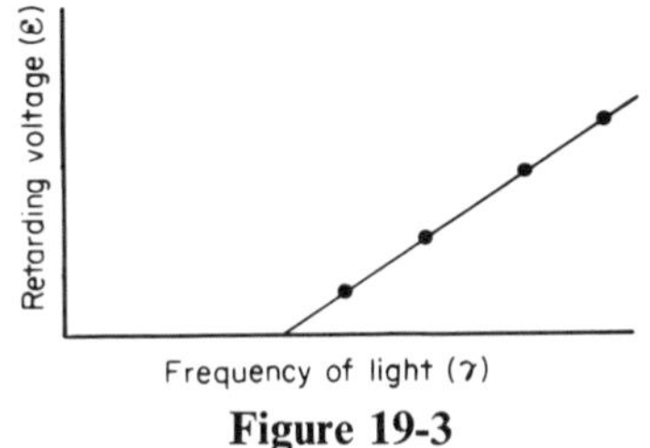

Figure 19-3

The straight line indicates that the frequency is proportional to the retarding voltage. Since wave length times frequency equals velocity and velocity is constant, (the speed of light in a vacuum is always 186,000 miles/sec for any wave length), it can be seen that the wave length is inversely proportional to the frequency.

Let us assume that the first plate was zinc. Now we will use cesium, and we might find that we get a current at 0 voltage for yellow light, a smaller frequency than blue light. Repeating the experiment now with different retarding voltages and different frequencies, we get a new line on our graph, and the graph looks like Figure 19-4.

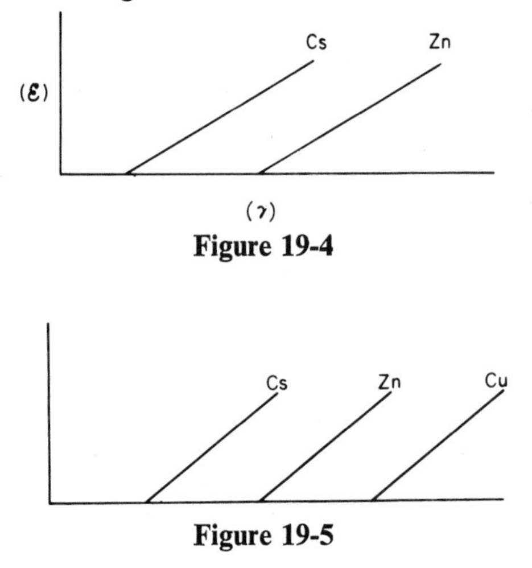

Figure 19-4

Figure 19-5

We do the same for copper, and a new line is added as in Figure 19-5.

It is obvious that in all cases the retarding voltage is proportional to the frequency of light and that the slope, or constant of proportionality, is the same in each case. This constant of proportionality is called Planck's Constant (h). Since the retarding voltage measures the energy of the emitted electrons, we can write an equation as follows:

Energy $= h\nu$

A quantum of light is called a photon. Please note that at no time do we state that light is a wave or a particle. We can only point out that experiments show that when we discuss the propagation of light through space, it appears as though light is behaving like a wave; and when we interpret the manner in which light reacts with matter, a particle model of light appears to be more satisfactory. We can, therefore, make no dogmatic statement regarding the exact thing that light is. We can only predict its behavior.

After studying Figure 19-5, we can extrapolate it below the zero energy level and draw it as in Figure 19-6.

One can then ask for the significance of the $(-E)$. Usually one of your students will come through for you and suggest that this represents the

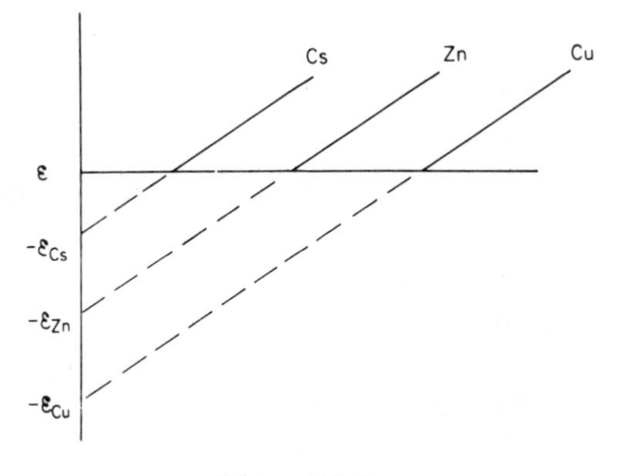

Figure 19-6

binding energy or the amount of energy the outer electron on this atom must absorb in order to overcome its potential and be torn loose. It reinforces the convention of using negative values for the energy of a bound electron.

These simple arguments augmented by the appropriate films and demonstrations are a good introduction to the subject, but will be much more palatable to your physics students who have been preparing all year to be introduced to this material. Your earlier work with waves in general (via water waves, sound waves, etc.) has prepared them for what will be discussed now.

It is interesting to point out that we are rediscovering a property of nature we had begun to see develop early in viewing the atomistic theory of matter, the graininess of charge, etc. Here again is evidence of graininess in nature; this time we see the graininess in energy. Of course, the grains are pretty small, Planck's constant has a value of 6.64×10^{-34} joule-sec., and so it takes some pretty sensitive experiments to sense this graininess. But then, we ran into this idea before in discussing the Millikan experiment and the discreteness of charge.

It is useful to assign enough problems for homework and subsequent discussion so that your students become familiar with and work easily with the relationships.

$$E = h\nu$$

and since $c = \nu\lambda$,

$$E = \frac{hc}{\lambda}$$

In a subsequent chapter, we will discuss special relativity. However, $E = mc^2$ is general knowledge and if $E/c = mc$, mc is momentum.

Therefore, if $E = \dfrac{hc}{\lambda}$

$E/c = h/\lambda$ and the momentum of a photon of given wave length (λ) is expressed $p = h/\lambda$, or in terms of frequency, $p = h\nu/c$.

Incidentally, *mks* units like meters and joules are incongruous in expressing properties of photons or discussing what goes on inside the atom. Just as cats to elephants can be easily measured in small numbers of the unit meter, atomic distances are all measured in small numbers of 10^{-10} meters. This measurement is called an Angstrom. We have already pointed out that the energy accumulated by one electron accelerated through a potential of one volt is 1.6×10^{-19} joules or one electron volt, abbreviated e.v.

Since $E = hc/\lambda$ and h and c are constants, we can develop a useful formula which can make problem solving much easier.

$$h = 6.6 \times 10^{-34} \text{ joule-sec and}$$

$$c = 3 \times 10^{8} \text{ m/sec}$$

$$E = \frac{6.6 \times 10^{-34} \text{ joule-sec}}{\lambda} \times 3 \times 10^{8} \text{ m/sec}$$

$$E = \frac{19.8 \times 10^{-20}}{\lambda} \text{ joule-meters}$$

$$E = \frac{19.8 \times 10^{20}}{\lambda} \text{ joule-meter} \times \frac{\text{ev}}{1.6 \times 10^{-19} \text{ joules}} \times \frac{A}{10^{-10} \text{ meters}}$$

$$E\lambda = \frac{19.8 \times 10^{-26}}{1.6 \times 10^{-29}} \text{ ev} - A = 12.4 \times 10^{3} \text{ ev} - A.$$

You can now refer to this as the "handy dandy" formula and permit your students to use it to convert photons of known wave length into their equivalent in energy, or vice versa.

Pressure of Light

Having developed a relationship for the momentum of photons, and having learned about the conservation of momentum in the study of Newtonion dynamics, your students can be asked whether or not they have ever seen evidence of this momentum of light. How would they look for it?

Since $F\Delta t = m\Delta v$, and $F = m\Delta v/\Delta t$, light reflecting from a surface should exert a force on that surface, and the force exerted over an area of that surface should be exhibited as a pressure (Pressure = Force/Area). A simple calculation can show that while this pressure might be present, it would be very hard to detect. For instance, a 100 watt lamp emits yellow light at a rate of 100 joules/sec. Suppose you enclosed this bulb in a sphere of one meter radius. What would the pressure on the walls of this sphere be? Let's assume an average wave length of about 5000 Angstroms. Then $E =$ 12400/5000 = 2.5 ev per photon or 2.5 (ev/photon) $\times$ 1.6 $\times$ 10^{-19} joules/ev = 4 $\times$ 10^{-19} joules/photon at a rate of (100 joules/sec)/(4 $\times$ 10^{-19} joules/photon) = 25 $\times$ 10^{19} photons/sec. Each photon is carrying a momentum $p = h/\lambda$ = (6.6 $\times$ 10^{34} joules/sec.)/5 $\times$ 10^{-9} meters and p = 1.3 $\times$ 10^{-27} newton-sec. If the wall is a black totally absorbing wall, then the force exerted on the wall is 1.3 $\times$ 10^{-27} (newton-second)/second or 1.3 $\times$ 10^{-27} newtons. If the wall is a totally reflecting silvered wall, then the momentum changes would be twice as much and the force exerted by the light on the walls of the sphere 2.6 $\times$ 10^{-27} newtons. The pressure would be Force/Area = (2.6 $\times$ 10^{-27} newtons)/$4\pi r^2$ = (2.6 $\times$ 10^{-27} newtons)/(4 $\times$ 3.14 $\times$ 1^2 meters^2) = (2.6 $\times$ 10^{-27} newtons)/(12.56 meters^2) = .27 $\times$ 10^{-27} newtons/m^2 or 2.7 $\times$ 10^{-28} newtons/m^2. This is truly a small pressure and would be very difficult to detect. (Sensitive instruments have detected the pressure of light, and cosmological phenomena like the direction of the tails of comets and the size of stars are explained by it.)

Compton Effect

The question might then arise—what happens when a photon is absorbed and reemitted in some different direction? If a photon can be said to carry momentum and we know that momentum is a vector quantity, then in order to conserve momentum the absorbing agent should recoil, and if it does it should carry away some energy. Can we detect this loss of energy in the reflected photon? This was a question proposed and demonstrated by Arthur Compton around 1923, for which he received a Nobel prize in physics.

In order for a perceptible amount of energy to be carried away in a collision, the mass of the target particle ought to be as close to the mass of the projectile as possible. We will use as a target the smallest particle we know and hit it with a photon of yellow light of 5000 A° wave length.

Let us assume that the photon strikes a free electron at rest and is reflected away at a path at right angles from its original path. The law of reflection still holds since both the angle of incidence and the angle of reflection are 45°.

In our everyday experience when we shine a yellow light on a metallic surface, yellow light is reflected from it and we can, therefore, draw a vector diagram of the collision as in Figure 19-7.

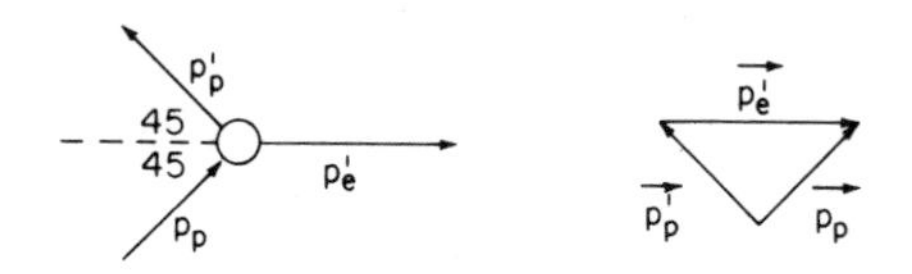

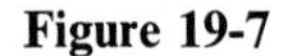

Figure 19-7

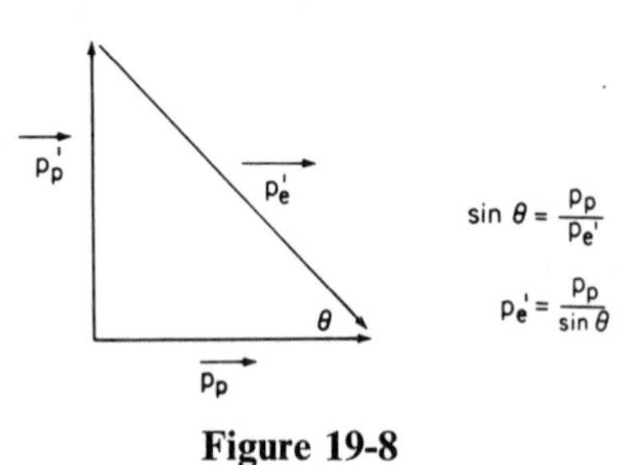

Figure 19-8

In order for the photon to come in and be reflected, and still conserve momentum, $\vec{p}_p = \vec{p}_p^{\,1} + \vec{p}_e^{\,1}$. (See Figure 19-8.)

The recoil momentum of the electron can then be seen to be

$$p_p = \frac{h}{\lambda} = \frac{6.63 \times 10^{-34}\ \text{joule-sec}}{5 \times 10^{-7}\ \text{meter}} = 1.3 \times 10^{-27}\ \text{newton-sec}$$

$$p_e^1 = \frac{1.3 \times 10^{-27}}{.707}\ \text{newton-sec} = 1.84 \times 10^{-27}\ \text{newton-sec}$$

The E_k of the electron after the collision is:

$$E_e = \frac{p^2}{2m} = \frac{(1.8 \times 10^{-27})^2}{2 \times 9.1 \times 10^{-31}\ \text{kg}} = \frac{3.4 \times 10^{-54}\ \text{joules}}{18.2 \times 10^{-31}}$$

$$E_e = 3.7 \times 10^{-24}\ \text{joules} \times \frac{1\ \text{ev}}{1.6 \times 10^{-19}\ \text{joules}} = 2.3 \times 10^{-5}\ \text{ev}$$

The energy of the original photon was

$$\frac{12400}{5000} \approx 2.5\ \text{ev}$$

Since 2.5 ev $-$ 2.3×10^{-5} ev $\approx$ 2.5 ev, it is evident that we could not detect the Compton effect in this collision, and the "common sense" approach of classical physics is not disturbed.

The retention of energy by the scattered photon is most probable with low energy photons since the energy supplied is not enough to remove the electron from the atom to which it is bound. If the entire atom must recoil, the comparatively immense mass of the atom would permit it to accept very little energy.

Let us, however, consider the same type of collision, only this time we will permit a much more energetic photon to collide with a free electron at rest.

Suppose we had a deep ultra-violet or soft X-ray photon of about 100 A° strike the electron and be scattered off at 90° as we had in the previous example. Would we be correct in assuming that the magnitude of the incoming and outgoing photon was the same (as close as could be measured) as we did with the 5,000 A° photon?

Let us figure it out.

$$p_p = \frac{h}{\lambda} = \frac{6.63 \times 10^{-34}\text{ joule-sec}}{10^{-8}\text{ meters}} = 6.63 \times 10^{-26}\text{ newton-sec}$$

Assuming again the same vector diagram as we had previously (it might be wrong in this case).

$$p_e^1 = \frac{6.63 \times 10^{-26}}{.707}\text{ newton-sec} = 9.38 \times 10^{-26}\text{ newton-sec}$$

$$E_e = \frac{p^2}{2m} = \frac{(9.38 \times 10^{-26})^2}{18.2 \times 10^{-31}} = \frac{88 \times 10^{-52}}{18.2 \times 10^{-31}} = 4.8 \times 10^{-21}\text{ joules}$$

$$E_e = \frac{4.8 \times 10^{-21}\text{ joules}}{1.6 \times 10^{-19}\text{ joules/ev}} = 3 \times 10^{-2}\text{ or ev}$$

The energy of the original photon was

$$\frac{12400}{100} = 124\text{ ev}$$

$124 - 3 \times 10^{-2}$ is still ≈ 124. While we are getting closer, it still appears classical.

Let us try a 1 A° photon, then

$$p_p = \frac{h}{\lambda} = \frac{6.63 \times 10^{-34}}{10^{-10}} = 6.63 \times 10^{-24}\text{ newton-sec}$$

and

$$p_e^1 = \frac{6.63 \times 10^{-24}}{.707} = 9.38 \times 10^{-24}\text{ newton-sec}$$

$$E_e = \frac{p^2}{2m} = \frac{(9.38 \times 10^{-24})^2}{18.2 \times 10^{-31}} = \frac{88 \times 10^{-48}}{18.2 \times 10^{-31}} = 4.8 \times 10^{-17}\text{ joules}$$

$$E_e = \frac{4.8 \times 10^{-17}\text{ joules}}{1.6 \times 10^{-19}\text{ joules/ev}} = 3 \times 10^{2}\text{ ev}$$

The energy of the photon was

$$\frac{12400A\text{-ev}}{1A} = 12400 \text{ ev}$$

$12400 - 300 = 12100$ ev, a measurable change in the energy of the photon, and since $E = h\nu$ or hc/λ, there is a measurable change in the frequency and the wave length of the scattered photon.

Several new ideas suggest themselves:

1. We can no longer think merely of a photon striking an electron and being reflected. The photon is absorbed by the electron, and then another photon is emitted as the electron radiates away its absorbed energy.
2. Most important, our procedure for solving the problem above was almost wrong since the vector diagram shown was based on an isosceles triangle with the momentum of the scattered photon equal in magnitude to the momentum of the incoming photon. Since $p = h/\lambda$ and λ is inversely proportional to $E = hc$, it is evident that if the energy of the scattered photon is less, its wave length will be longer and its momentum will decrease. In this case the change is still small enough so that the figure is approximately correct.

To take the change in the momentum of the scattered photon into account, we need to evolve a different technique for solving such a problem. But when an electron scatters a photon with sufficient energy to create a strong Compton Effect, chances are the electron will recoil with a velocity close to the speed of light. It therefore becomes necessary to wait until we have looked into special relativity before we can discuss the Compton Effect in detail.

We have discussed the photo-electric effect and the Compton Effect. Other particle-photon reactions are worth bringing to the attention of your students.

X-Ray Production

When electrons boil off a hot filament at a high potential and are then suddenly brought to rest by striking the metal anode of a vacuum tube, the energy lost by the electron is radiated away. In ordinary electronic vacuum tubes where the potential is of the order of 100 volts, much of the energy is radiated away in the form of infra-red light, and possibly some ultra-violet. If, however, as in the case of the X-ray tube, the accelerating potential is in the order of 10^4 volts, then a 10,000 ev electron will strike the anode, come to rest quickly, and give rise to a 10,000 ev photon.

$12400/10000 = 1.24$ A°—a one Angstrom photon is certainly in the "hard" X-ray region.

Incidentally, it might be worth pointing out here that while we subdivide the electro-magnetic spectrum into classifications like visible light, ultra-violet, X-rays, Gamma rays, etc., these are by no means sharp divisions. The classifications of ultra-violet, X-rays and Gamma rays describe their origin more than their limiting wave lengths. For instance, ultra-violet radiation is the name given to the energetic radiation from excited electrons in an atom—these photons might have wave lengths down to 100 A°. X-ray is the name given to the emanations occurring when highly energetic electrons are stopped by striking a metal anode. These are artificially produced and range from about 1 A° to 100 A°, overlapping with ultra-violet in the "soft" X-ray region and with Gamma radiation in the "hard" X-ray region. Gamma Ray is the name given to the electro-magnetic radiation coming from excited nuclei.

Other forms of photon-electron interactions, like pair production and pair annihilation, might be mentioned in passing here but should be discussed in more detail with your students after studying the principles of relativity.

The natural extension made by Louis DeBroglie was that if electro-magnetic waves turn out to behave in a particle-like manner, then perhaps particles might behave in a wave-like manner.

Your students might be interested to know that DeBroglie offered this thesis as part of a dissertation for his PhD, and it might have been rejected by his professor had not Albert Einstein been visiting in Paris at the time. Dr. Einstein thought the idea had merit and that it was a worthy PhD thesis. Dr. DeBrogie received the Nobel prize in physics later for this contribution to modern physics. The idea is simple enough.

If a photon has a momentum of $p = h/\lambda$, then a particle (let us start with the smallest particle, like an electron) has a wave length $\lambda = h/mv$.

Germer and Davidson later checked out this idea by reflecting electrons from crystal surfaces (acting like diffraction gratings with $d \approx 1$ A°). The electrons which then impinged on a photographic plate made a pattern which resembled the interference pattern seen when light is reflected from a diffraction grating. The calculations made by measuring the distance between the fringes (x), the distance of the plate from the crystal surface (l), and the known geometry of the crystal lattice from which the parameter (d) is obtained gave the wave length $\lambda = x \cdot d/1$ (a wave length which agreed with DeBroglie's prediction that $\lambda = h/mv$). Germer and Davidson also received the Nobel prize in physics for their experimental verification of DeBroglie's thesis.

This wave-like behavior of particles may not, at first, fit in with a "common sense" approach to particles, and students sometimes try to fit it into their minds by talking about particles moving in wavy paths. Stop that at once. The duality here is somewhat akin to the aborigine we cited earlier who tried to describe an airplane to his fellows.

The wave-likeness of a particle best describes the probability of the particle's location. A demonstration I use to emphasize this point is done as follows:

I set up an inclined board on which marbles roll down, making a series of collisions with pegs set in the path, and accumulate at the bottom of the board in a predictable probability pattern. (See Figure 19-9.)

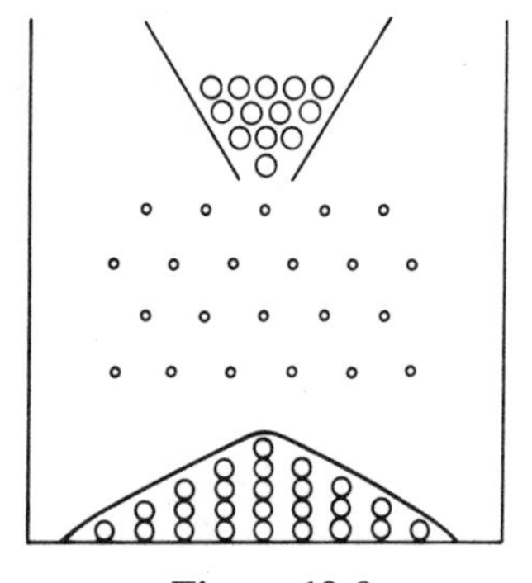

Figure 19-9

The central peak suggests that this is probably where one could most safely predict that a marble might be located after rolling down the board. Suppose, however, the pattern, after rolling many marbles down the board, looked like Figure 19-10. This indicates the probability that where a marble would roll is the greatest where the peaks are highest and smallest in the valleys. Yet it looks like an interference pattern which indicates a wave-likeness, but a wave-likeness not assignable to any single particle. It only describes the behavior of large numbers of particles—statistical numbers where probability becomes more certain.

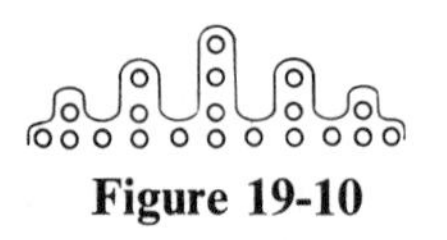

Figure 19-10

Some practical application of this matter wave principle might be cited. One will be discussed in some detail in discussing atomic structure where it can be shown that the electron wave length confirms Bohr's atomic model which he arrived at empirically from spectrographic data.

The electron microscope is a practical application of the wave nature of the electron. Students will remember that when waves move by an object which is smaller than the wave length, the waves are diffracted around the object and very little shadow is cast. When we wish to view small things in an ordinary visible light microscope, we find a limit to the magnification we can attempt. We reach a point where we begin to view details which are smaller than the wave length of the light we are using. It then becomes useless to continue magnifying because the more we magnify the fuzzier the outlines of our images become.

It then becomes useful to use smaller wave lengths (we might not see them but we can photograph them). However, we know that $E = hc/\lambda$, and as we decrease the wave length of our light for greater detail of small objects, we introduce photons of higher energy which might destroy and distort the object we are trying to view.

Let us compare the energies of a 10 A° X-ray photon and a 10 A° wave length electron.

$$E_{\text{(photon)}} = \frac{12400}{10} = 1{,}240 \text{ ev}$$

for electron

$$mv = \frac{h}{\lambda} = \frac{6.6 \times 10^{-34} \text{ joule-sec}}{10^{-9} \text{ meters}} = 6.6 \times 10^{-25} \text{ kg meter/sec}$$

$$E = \frac{p^2}{2m} = \frac{(6.6 \times 10^{-25})^2}{18.2 \times 10^{-31}} = \frac{40 \times 10^{-50}}{18 \times 10^{-31}} \approx 2 \times 10^{-19} \text{ joules}$$

$$E_{\text{(electron)}} = \frac{2 \times 10^{-19} \text{ joules}}{1.6 \times 10^{-19} \text{ joules/ev}} \approx 1 \text{ ev}$$

It is quite evident that electrons with equivalent wave lengths carry much less energy than X-rays. This is the reason electron microscopes are used when it is necessary to view details requiring extremely high magnification.

Another interesting application of matter waves is one way of looking at the phenomenon of the superfluidity of helium at very low temperatures. Helium remains liquid at temperatures very close to 0° Kelvin. However, at these extremely low temperatures, it exhibits the strange property of being uncontainable. That is, a small amount of liquified helium placed in a deep beaker appears to be able to climb up the wall of the beaker and outside when brought down to near zero Kelvin.

If, as we pointed out before, the wave length of a particle is the probability of its location in space and $\lambda = h/mv$, it becomes evident that at very low temperatures the momentum of the helium atom would be very small and its wave length becomes very long. If its wave length becomes longer than the dimensions of its container, then the probability of finding helium atoms outside the container, as well as inside, becomes greater and that is exactly what happens.

The expression $p \cdot \lambda = h$ is an expression of uncertainty. Actually, the Uncertainty Principle as expressed by Heisenberg and derived somewhat differently is $\Delta p \cdot \Delta x = \hbar$ where $\hbar = h/2\pi$, but the approximation is close enough. Perhaps the real definition of a particle is its ability to be located in space. A baseball with its large mass could be very cold indeed and still have sufficient momentum so that its wave length would be insignificant. For instance, suppose we consider a mass of 10^{-1} kg moving with a velocity of 1 meter per year or 10^{-7} meters per second, then its wave length would be $\lambda = h/mv = (6.6 \times 10^{-34}$ joule-sec$)/(10^{-1} \times 10^{-7}$ kg m/sec). λ, which is analogous to the Δx, the uncertainty of position in the Heisenberg Principle, would be 6.6×10^{-26} meters. Its particle-likeness would still be most apparent since it could be very easily located.

20

The Atom

All visible objects, man, are but as pasteboard masks.

HERMAN MELVILLE IN *Moby Dick*, CHAPT. 36.

WHAT WE PERCEIVE as reality may be a false front, a pasteboard reality to what lies deeper. Geiger's and Madsen's scattering experiments in which alpha particles were seen to be scattered through various angles by very thin gold foil and Rutherford's interpretation of their results is an excellent example of "model-building." Prior to these experiments performed about 1913, it was known that atoms consisted of positive and negative particles. J. J. Thomson had already shown in 1897 that electrons were particles. He measured the charge to mass ratios of electrons, but how electrons were distributed in the atom was unknown.

A natural model was the one in which electrons were pictured as distributed throughout some positively charged medium. The Rutherford model laid this to rest. Ask your students to interpret the experimental results. Start with this "plum-pudding model." In solids, atoms are pictured as being in contact with one another. Therefore, the alpha particles flying toward the thin gold foil would "see" an impregnable wall and they should all bounce back if the wall is hard. If the material of which the wall is made is soft, they should be all able to go through. But how can one explain what was observed?

The PSSC textbook draws the analogy by likening it to bullets fired at stacks of bales of hay and finding most of the bullets going straight through, some being deflected and some even being reflected back. Other analogies could be drawn, but somehow a picture must be painted of atoms consisting of very tiny, relatively massive nuclei in their centers containing all the positive charge while the tiny feathery electrons float about in the empty space of which most of the atom consists. It was an intuitive guess on Rutherford's part that led him to an inverse square law repulsive force from a very small but comparatively massive nucleus.

His guess that the nucleus was positive in order to repel the positively charged alpha particles fits in with subsequent investigations and agrees with our current model of the atom. It is interesting to note that to explain the observation of Geiger and Madsen, an attractive force applied by a negatively charged nucleus would have been just as satisfactory. (See Figures 20-1a and 20-1b.)

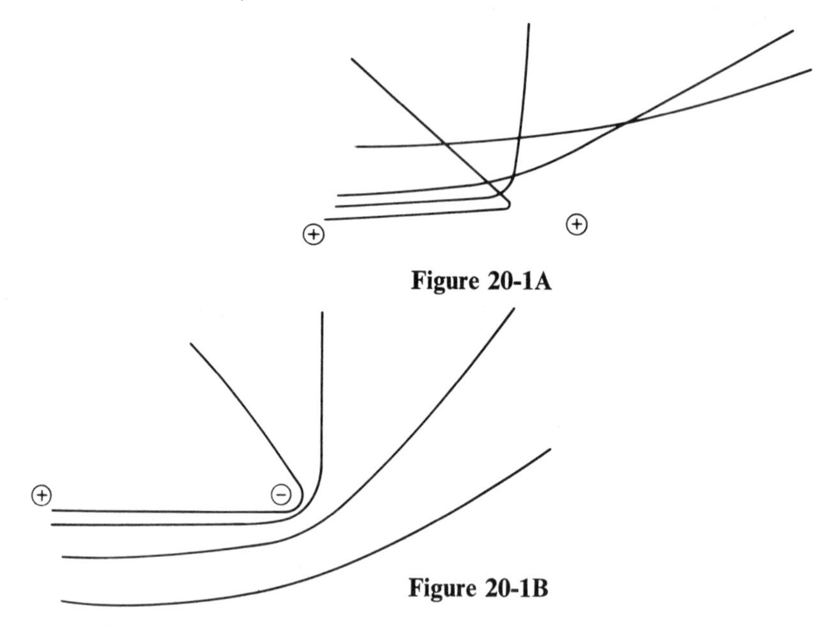

Figure 20-1A

Figure 20-1B

The model of a coulomb hill to climb as the alpha particle approaches the nucleus is easy to illustrate.

I gave one of our wood shop teachers an inverse square law graph and asked him to construct such a hill about ten inches high. The device is pictured in Figure 20-2.

A steel bearing can be launched down a ramp with various energies and different aiming errors with respect to the "nucleus" on top of the hill and the path can be observed. The device shown is useful in introducing some of the concepts needed in nuclear physics, since it can be seen that

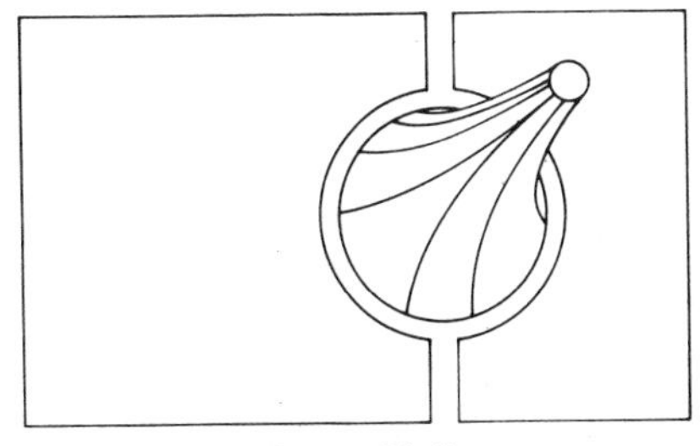

Figure 20-2

in order for a positively charged particle to reach the nucleus, it must have enough energy to climb all the way up the potential hill. The chance of reaching the nucleus is still very small since the cross-section of the nucleus is extremely small with respect to the diameter of the atom. A positive particle must, therefore, have a great deal of energy and must be heading directly for the nucleus in order to be captured by the nuclear forces.

From here, one can lead to discussions of the high energy particle accelerators used today in nuclear research. Illustrations suggested here can also lead to the presentation of "fusion" reactions and the necessity for achieving extremely high temperature plasmas (and the problems such high temperatures present) in order to drive the reaction to fusion.

Rutherford's model of a planetary system, like atoms with electrons orbiting a central nucleus, was a step forward and most appealing to the imagination. Some very simple questions stop complete acceptance of this model. Why is each hydrogen atom unidentifiable from any other hydrogen atom even though their past history must have been very different? Why are there only 92 natural elements? A simple planetary model would permit infinite variations.

Of course, a discrepancy which was most disconcerting was that according to classical electro-magnetic theory, stable atoms had no business existing at all. An electron moving in any path other than a straight line is an accelerating charge and an accelerating charge must radiate away energy, yet in the unexcited atom no energy can be lost. Also, how does one explain that an atom can be identified by the spectral lines it emits when it is excited. (Your students should, by this time, have observed excited gas tubes through diffraction gratings.)

Bohr's contribution here was the next giant step toward the acceptance of the atomic model as we see it today.

It was he who first recognized that Planck's constant (h), which Einstein had shown must play a part in explaining the photo-electric effect, must also be involved in explaining the discrepancies in the Rutherford model discussed above.

Let us do a dimensional analysis on Planck's Constant.

$$h = 6.63 \times 10^{-34} \text{ joule-sec}$$

$$\text{joule} \times \text{sec} = \text{kg}\,\frac{m^2}{\text{sec}^2} \times \text{sec} = \text{kg}\,\frac{m}{\text{sec}} \times m$$

Angular momentum, by definition, is $mv_{\perp} \times r$ and has, therefore, the same dimensions as kg m/sec $\times$ m. It was Bohr's "hunch" that it was the quantization of angular momentum that explained what classical physics could not then explain.

Assuming that the energy of emitted photons from excited atoms equals $h\nu$, the energy must be equal to the energy loss as an excited electron drops down from excited level to a lower level.

$$h\nu = Er_f - Er_i$$

The relationships derived by Bohr were based on spectroscopic evidence. While his derivations should be historically interesting, I would suggest that we can reach the same conclusions he came to by using the wave nature of the electron as proposed by DeBroglie.

In order for an electron to remain in a stable orbit, its wave nature predicts that the circumference of the orbit must be a whole number (n) of its wave length (λ)

$$n\lambda = 2\pi r$$

According to the DeBroglie principle,

$$\lambda = \frac{h}{mv}$$

so

$$\frac{nh}{mv} = 2\pi r$$

and

$$n\,\frac{h}{2\pi} = mvr$$

Since mvr has been defined as the angular momentum, we can see that the angular momentum is quantized in whole number units of $h/2\pi$. This constant $h/2\pi$ is called h-bar and represented by the sign $\hbar$.

We can develop the Rydberg formulas for the energy and radius of a hydrogen atom with several simple logical steps. Remember, we will use the DeBroglie relationship. Impress on your students that what we are doing deductively was done by Bohr and Rydberg, Balmer and others by fitting together experimental observations.

We start with $\dfrac{nh}{mv} = 2\pi r$

From classical dynamics for circular motion, we know that

$\dfrac{mv^2}{r} = \dfrac{kqZ}{r^2}$ and $mv^2 = \dfrac{kqZ}{r}$

where m is the mass of the rotating electron
v is its speed
k is the coulomb constant
q is the charge on the electron
Z is the charge on the nucleus
r is the radius of the electron's orbit

but
q is one elementary charge

so

$$mv^2 = \frac{kZ}{r}$$

Remember that the potential energy (u) of a bound electron is $-kZ/r$ and the total energy $E = u + \frac{1}{2}mv^2$

$$\text{or} \quad E = -\frac{kZ}{r} + \frac{kZ}{2r} = -\frac{kZ}{2r}$$

We can multiply our first two equations together.

$$\frac{nh}{mv} \times mv^2 = 2\pi r \times \frac{kZ}{r}$$

to obtain $nhv = 2\pi kZ$
and therefore v (the speed of the orbiting electron) $= 2\pi kZ/nh$
substituting into

$$\frac{nh}{mv} = 2\pi r$$

we get

$$\frac{n^2h^2}{m2\pi kZ} = 2\pi r$$

$$\frac{n^2h^2}{mkZ} = 4\pi^2 r$$

and

$$r = \frac{n^2h^2}{4\pi^2\, mkZ}$$

Since

$$E = -\frac{kZ}{2r}$$

then

$$E = -\frac{kZ}{\dfrac{2n^2h^2}{4\pi^2mkZ}} = -\frac{2\pi^2k^2m}{h^2} \cdot \frac{Z^2}{n^2}$$

$-\dfrac{2\pi^2k^2m}{h^2}$ is a constant whose value we can determine. (We will call this constant R.)

$$R = \frac{2\pi^2k^2m}{h^2} = \frac{2 \times (3.14)^2 \times (2.3 \times 10^{-28})^2 \times 9.1 \times 10^{-11}}{(6.63 \times 10^{-34})^2}$$

$$= \frac{2 \times 9.87 \times 5.3 \times 10^{-56} \times 9.1 \times 10^{-31}}{44 \times 10^{-68}}$$

$$= 21.8 \times 10^{-19} \text{ joules}$$

$$21.8 \times 10^{-19} \text{ joules} \times \frac{\text{ev}}{1.6 \times 10^{-19} \text{ joules}} = 13.6$$

Therefore $E_n = -R(Z^2/n^2)$ and $R = 13.6$

R is called the Rydberg constant having been determined by Rydberg from data accumulated by spectroscopy experiments.

Note that when we plug in $Z = 1$ and $n = 1$ (the values for a hydrogen atom at rest) we get -13.6 ev, and 13.6 ev or 21.8×10^{-19} joules is exactly the amount of energy required to ionize a hydrogen atom from its rest state.

Since

$$E = -\frac{kZ}{2r}$$

$$r_n = -\frac{kZ}{2E_n} = \frac{-kZ}{-2R(Z^2/n^2)} = \frac{kn^2}{2RZ}$$

again when

$Z = 1$ and $n = 1$

$$r = \frac{k}{2R} = \frac{2.3 \times 10^{-28}}{2 \times 21.8 \times 10^{-19}} = \frac{2.3 \times 10^{-28}}{43.6 \times 10^{-19}}$$

$$r = \frac{2.3 \times 10^{-28}}{4.4 \times 10^{-18}} = .52 \times 10^{-10} \text{ meter or .52 angstrom}$$

This value (.52 A°) is the approximate radius of a hydrogen atom. Since $r_n = -kZ/2E_n$ and $-k/2E_n = .52$

then $r_n = .52\, Z/E_n$ and we can use this relationship to determine the radius of other atoms that behave like hydrogen.

Why would we not expect other atoms to behave exactly like hydrogen?

In deriving the Rydberg formulas we started with a simple coulomb force existing between the outer electron and the charge in the nucleus, but the situation is not so simple in multi-charge atoms where electrons in inner shells will screen out some of the nuclear charge so the Z effective is not equal to Z.

Multiple electron atoms behave in a hydrogen-like manner, therefore, when the atom is partially ionized so that the nucleus is not shielded (as in singly ionized helium or doubly-ionized lithium).

An excited electron in an un-ionized atom might also go through hydrogen-like energy level jumps when n is large since, if it were far enough out, it could "see" the nucleus and the inner electrons as a single particle. Z would then be simply p (the number of protons in the nucleus), minus e (the number of electrons close to the nucleus and between the nucleus and the distant excited electron).

Pauli Exclusion Principle

Such relationships coupled with spectroscopic data led early investigators to realize that whereas in H and He the rest state for the electrons was where $n = 1$, in lithium two electrons went down to the $n = 1$ state while one remained at the $n = 2$ state. Detailed analysis led to the construction of a system of electron shells which was empirically justifiable but for which there was no explanation.

Pauli pointed out that besides the n-state of the electron being quantized, its orbital angular momentum (l), its orbital magnetic moment (m_e) and its spin magnetic moment (m_s) are also quantized and no two electrons in an atom can have all four quantum numbers (n, l, m_e, m_s) the same; or no two electrons in an atom can exist in the same state.

It thus became possible to justify the number of electrons which could exist in each n shell. It might be further noted that these different quantum numbers explain "fine structure" spectroscopy. For instance, an electron with its magnetic momentum lined up with the nuclear magnetic moment would have a lower energy potential than one with its spin magnetic moment oriented opposite to the nuclear magnetic moment.

Your students might be interested in knowing that Pauli is supposed to have thought of this explanation for the anomalies bothering spectroscopists while he was attending an opera with his wife. It indicates that a boring opera was not all bad—it led to a major contribution in physics.

A PROBLEM AND LAB EXERCISE

A good exercise for students to implement the concepts herein presented follows:

Set up several hydrogen discharge tubes around the laboratory and have your students measure the wave lengths of the emitted spectra as accurately as possible. Using meter sticks and replica grating as suggested in Chapter 9 would be satisfactory if nothing better is available. It would, however, require extremely good technique to get precise measurements. If spectroscopes are available, use them. The spectroscope described for use with the PSSC Advanced Topics course is a good, yet inexpensive, one.

Using the relationship derived early, $E = -R(Z^2/n^2)$, have your students construct an energy level diagram for hydrogen.

Remember $R = 13.6$ and Z for hydrogen $= 1$.

Therefore, for $n = 1$, $E = -13.6 \cdot \frac{1}{1} = -13.60$ ev
for $n = 2$, $E = -13.6 \cdot \frac{1}{4} = -3.40$ ev
for $n = 3$, $E = -13.6 \cdot \frac{1}{9} = -1.51$ ev
for $n = 4$, $E = -13.6 \cdot \frac{1}{16} = -0.85$ ev
for $n = 5$, $E = -13.6 \cdot \frac{1}{25} = -0.54$ ev

The derived energy level diagram would then look like Figure 20-3.

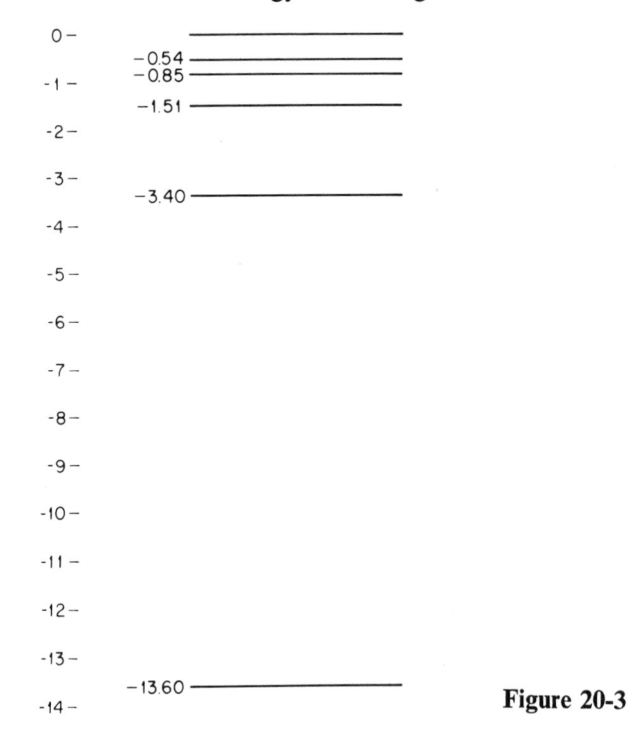

Figure 20-3

Students would then be asked to fit the observed and measured wave lengths into the derived energy level diagram. They can use the previously derived "handy-dandy formula," $E = 12400/\lambda$ A° ev. It is evident from inspection of their energy level diagram that the visible light would entail drops to the $n = 2$ level, since a drop from $n = 2$ to $n = 1$ calls for 10.2 ev which would be in the ultra-violet region, and a maximum energy photon emitted by a drop to the $n = 3$ level would be 1.51 ev or $\lambda = 8200$ A° (infra-red light).

There should be a great deal of satisfaction in observing that experimentally measured energy drops fit so exactly between lines within the predicted limits.

An experiment suggested in the PSSC Advanced Topics uses similar techniques but suggests that knowing that these observed energy emissions are $n = 2$ transitions and that since

$$E_n = -R \cdot \frac{Z^2}{n^2} = -\frac{2\pi^2 k^2 m}{h^2} \cdot \frac{1}{n^2} \text{ and}$$

$$E_{n_f} - E_{n_i} = -\frac{2\pi^2 k^2 m}{h^2}\left(\frac{1}{n_f^2} - \frac{1}{n_i^2}\right) \text{ since}$$

$$\Delta E = h\nu = E_{n_f} - E_{n_i}$$

$$\nu = -\frac{2\pi^2 k^2 m}{h^3}\left(\frac{1}{n_f^2} - \frac{1}{n_i^2}\right)$$

but

$$\nu = \frac{c}{\lambda} \text{ or } \frac{\nu}{c} = \frac{1}{\lambda}$$

therefore,

$$\frac{1}{\lambda} = \frac{2\pi^2 k^2 m}{h^3 c}\left(\frac{1}{n_f^2} - \frac{1}{n_i^2}\right)$$

Since λ is measured, n_f and n_i can be guessed, k, m, and c are known, h can be derived with considerable accuracy. It is a worthwhile experiment.

21

Special Relativity

0, it is excellent to have a giant's strength; but it is tyrannous to use it like a giant.

SHAKESPEARE, *Measure for Measure*
Act II, Scene ii

THE ACCEPTANCE of the wave nature of light in the nineteenth century led to the question, "what's waving?" and thus to the invention of ether as a carrier for these waves. The search for the experimental evidence of the existence of this ether led Michelson and Morley to perform an experiment in 1887.

The Search for an Ether "Wind"

In Chapter 4, we found that if a boat made a trip on a river with a velocity (v) with respect to the water, while the water moved with a velocity (u) with respect to the land, the time to make the trip downstream and back would be $T_{//} = (T_1)/(1 - v^2/c^2)$ where $T_{//}$ is the time required to make the trip in moving water parallel to the current and T_1 is the time required to make the trip in still water.

Would the time to go back and forth across the river perpendicular to the current be any different? (See Figure 21-1.)

The boat would have to be pointed partly upstream on both the crossing to the opposite bank and return. The time ($T_{\perp}$) to make the crossing is

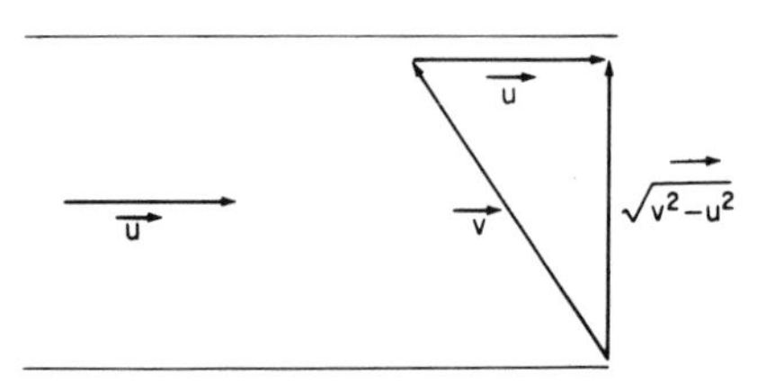

Figure 21-1

therefore, $T_{\perp} = \dfrac{d}{\sqrt{v^2 - \mu^2}}$, when d is the back and forth distance across the river.

$$T_{\perp} = \frac{d/v}{\sqrt{v^2 - \mu^2/v^2}} = \frac{d/v}{\sqrt{1 - (\mu^2/v^2)}}$$

but d/v is T_1, the time to make the trip in still water, so

$$T_{\perp} = \frac{T_1}{\sqrt{1 - (\mu^2/v^2)}}$$

Note that $T_{//}$ and $T_{\perp}$ are not the same.

It was Michelson's and Morley's scheme to use this difference in time travel, parallel to the stream and perpendicular to the stream, to test for the presence of an ether.

They reasoned that as the earth moved through space around the sun through the ether, an ether wind or current should be present and should be able to be experimentally detected.

In order to find the very small difference in time they expected, Michelson invented an interferometer which is schematically sketched in Figure 21-2.

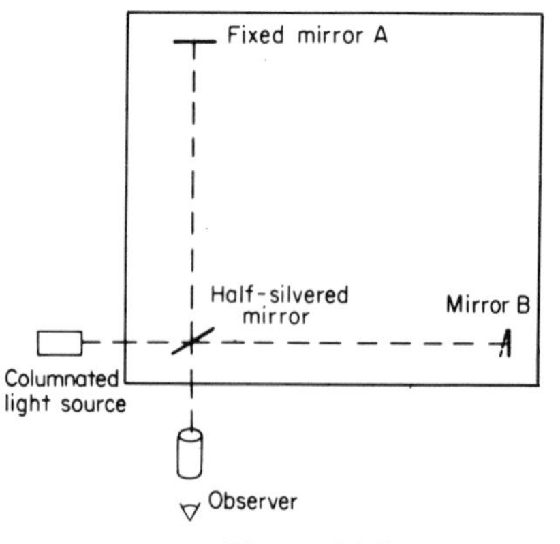

Figure 21-2

Mirror A is fixed perpendicular to the path of incident light so the affected beam goes straight back. Mirror B is set at a very slight angle to the incident light which reaches it after going through the half-silvered mirror. The two reflected light beams which are recombined and converged at the observation point will have varying path difference along the cross-section of the beam and will show interference fringes.

This, in fact, does happen and can be easily demonstrated with such a piece of equipment. The Michelson interferometer designed for a PSSC Advanced Topics laboratory is an excellent and relatively inexpensive device.

Michelson and Morley mounted their interferometer on a large stone (like an old fashioned mill-wheel) which they then floated in a pool of mercury. This was done to minimize disturbing vibrations. They could then slowly rotate the entire apparatus as they observed the fringes. On a 180° degree turn, the light path to mirror (A) might be parallel to the direction of motion of the earth, then both light paths would be 45° to the earth's motion and, finally, path (A) would be perpendicular while the path to mirror (B) would be parallel to the velocity of the earth. As a result, it was reasoned that a fringe shift would be seen, as the time for the light to go from the source to the mirror and back to the observer would change. This did not happen. Meticulous observations indicated that there was no change in the speed of light. Morley's colleagues commiserated with him on the "failure" of his experiment. It was a magnificent failure. It made physicists rethink of the requirement for an ether and it strongly suggests that since such a sensitive experiment could not detect an ether "wind," perhaps no ether actually exists.

Though Einstein, in his later writings, reports that he had not known of the Michelson-Morley experiment prior to the publication of his Theory of Special Relativity, it is likely that it was the effect of this experiment, plus consideration of other anomalies, which led Albert Einstein to state that the speed of light in a vacuum was constant and was independent of the observer or the source.

Special Relativity—Time Dilation

Following are some interesting and unusual derivations of special relativity for presentation to high school students. Since I plan to use the format of a "gedanken" experiment with a photon clock, it becomes necessary to develop time dilation first in order to use these concepts for the development of space relationships in inertial frames that are moving with respect to one another.

For our clocks, to compare intervals of time for a stationary observer and a moving assistant (henceforth called the rider), we will use a vertical tube in which a photon rises from the bottom to the top. As the rider goes by the observer with a velocity (v), both carry their photon tubes vertically and the photons in both are emitted from the bottom as they go by each other. After a time Δt_0, as measured by the observer's clock, the photon has risen to the top of his tube and his tube has a length $l_0 = c\Delta t_0$. The assistant, at rest with respect to his tube, thinks the same about his clock, but to the observer, the rider's photon has traveled the longer diagonal path as seen in Figure 21-3.

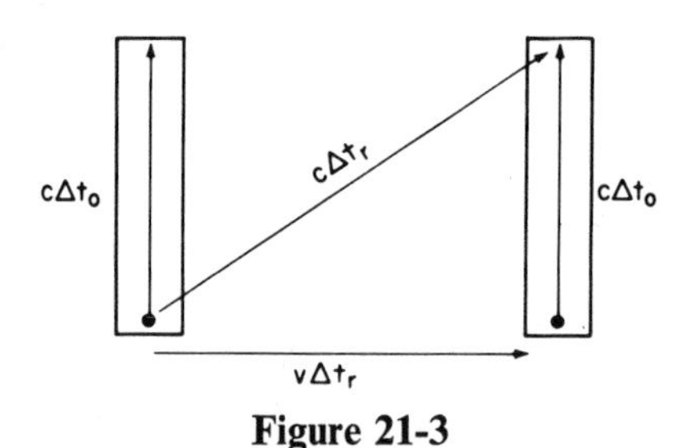

Figure 21-3

Since c must be constant, Δtr (the duration of time for the rider's photon to reach the end of his tube as measured by the observer's clock) must be greater than Δt_0. $c\Delta tr$ is the length of the path the rider's photon took while the rider moved a distance of $v\Delta tr$. Therefore, by the Pythagorean relationship

$$c^2\Delta t_r^2 = v^2\Delta t_r^2 + c^2\Delta t_0^2$$
$$c^2\Delta t_0^2 = c^2\Delta t_r^2 - v^2\Delta t_r^2$$
$$c^2\Delta t_0^2 = \Delta t_r^2(c^2 - v^2)$$
$$\Delta t_0^2 = \Delta t_r^2\left(1 - \frac{v^2}{c^2}\right)$$
$$\Delta t_0 = \Delta t_r\sqrt{1 - \frac{v^2}{c^2}}$$
$$\Delta t_r = \frac{\Delta t_0}{\sqrt{1 - v^2/c^2}} \qquad \textbf{(Equation no. 1)}$$

Space Contraction

Now the observer decides he is ready to compare the distance measurements and he calls his assistant, the rider, back to make the necessary adjustments and start over again.

He installs mirrors at the ends of the rider's photon tube so that an emitted photon will travel to the opposite end of the tube and back again. He also has the traveler lay his tube on its side, since he must compare horizontal distance along path.

The rider starts his trip, and when he reaches a constant velocity $\vec{v}$, he causes a photon to be emitted in his (to him) stationary horizontal tube and observes that the time (Δt_0) it takes for the photon to go to the mirror and back to the source is $\Delta t_0 = 2x_0/c$ where x_0 is the length of the tube in Figure 21-4.

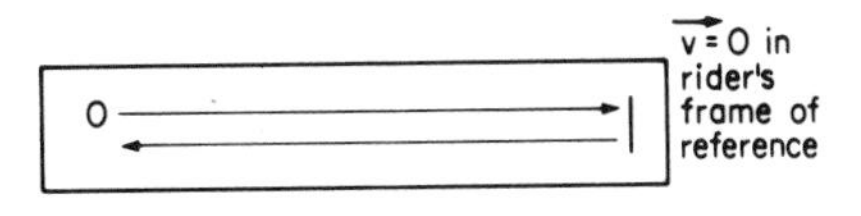

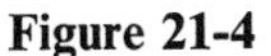

Figure 21-4

To the observer, however, the rider's photon moves up the tube a distance of $x_r = c\Delta t_{r_1} + v\Delta t_{r_2}$ and moves back a distance of $x_r = c\Delta t_{r_2} - v\Delta t_{r_2}$. (See Figure 21-5.)

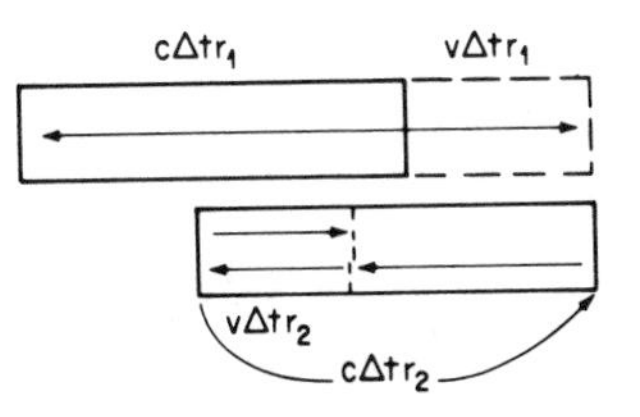

Figure 21-5

$$\Delta t_{r_1} = \frac{x_r}{c+v}, \quad \Delta t_r = \Delta t_{r_1} + \Delta t_{r_2} = \frac{x_r}{c+v} + \frac{x_r}{c-v}$$

$$\Delta t_{r_2} = \frac{x_r}{c-v}$$

$$\Delta t_r = \frac{x_r(c-v) + x_r(c+v)}{c^2 - v^2} = \frac{2x_r c/c^2}{(c^2 - v^2)/c^2}$$

$$\Delta t_r = \frac{2x_r/c}{1 - v^2/c^2} \text{ but } \Delta t_r = \frac{\Delta t_0}{\sqrt{1 - v^2/c^2}}$$

$$\therefore \frac{\Delta t_o}{\sqrt{1 - v^2/c^2}} = \frac{2(x_r)/c}{1 - v^2/c^2}$$

$$\Delta t_0 = \frac{2x_r/c\sqrt{1 - v^2/c^2}}{1 - v^2/c^2}$$

$$\Delta t_0 = \frac{2x_r/c}{\sqrt{1 - v^2/c^2}}$$

$$\text{but } \Delta t_0 = \frac{2x_0}{c}$$

$$\therefore \frac{2x_0}{c} = \frac{2x_r/c}{\sqrt{1 - v^2/c^2}}$$

$$x_0 = \frac{x_r}{\sqrt{1 - v^2/c^2}}$$

$$x_r = x_0 \sqrt{1 - \frac{v^2}{c^2}} \qquad \textbf{(Equation no. 2)}$$

Equivalence of Mass and Energy

Since the speed of light is the ultimate speed, what happens to the energy one adds to a particle as it approaches the speed of light? Certainly its kinetic energy as we usually express it ($\frac{1}{2} mv^2$) cannot keep increasing indefinitely since the velocity v is limited by c, the speed of light.

Experiments show that the velocity of particles does not increase as energy is added. An excellent illustration of this is a filmed experiment called "The Ultimate Speed" made by Education Services, Inc. and distributed for sale or rented by "Modern Learning Aids."

The fact that $E = mc^2$ is further illustrated in the discovery of a fission reaction by Hahn and Strassmann, which led through the agencies of Lisa Meitner, Neils Bohr and Enrico Fermi to the destruction of Nagasaki and Hiroshima and the end of the second World War.

The book *Atoms in the Family* by Laura Fermi is recommended reading. The story of the Manhattan Project from the Hahn and Strassmann discovery to the test at Los Alamos is therein depicted and it rivals any "cloak and dagger" sequences in fiction.

The incorporation of the tremendous amount of energy incorporated in a relatively small amount of mass might be illustrated by the following examples:

Let us assume we had the means of converting one kilogram (about 2.2 lbs.) of sugar into pure energy, what would it amount to? If $E = mc^2$, then $E = 1 \text{ kg} \times (3 \times 10^8 \; m/\text{sec})^2$.

$$E = 9 \times 10^{16} \frac{\text{kg } m^2}{\text{sec}^2} \text{ or } 9 \times 10^{16} \text{ joules}$$

1 watt = 1 joule/sec.

Our cost for electricity in most places is approximately two cents per kilowatt-hour.

$$1 \text{ kilowatt-hour} = 1000 \frac{\text{joules}}{\text{sec}} \times 3600 \text{ sec.} = 3.6 \times 10^6 \text{ joules}$$

$$\frac{9 \times 10^{16} \text{ joules}}{3.6 \times 10^6 \dfrac{\text{joules}}{\text{kilowatt-hr.}}} = 2.5 \times 10^{10} \text{ kilowatt-hours}$$

$$2.5 \times 10^{10} \text{ kilowatt-hrs.} \times \frac{2 \times 10^{-2} \text{ dollars}}{\text{kilowatt-hr.}} = 5 \times 10^8 \text{ dollars}$$

About two lbs. of material would yield about 500 million dollars worth of electrical energy, if all the mass could be converted into energy.

Bombs stocked today are in the megaton range or better. Let us try to help our student visualize what we mean by an energy yield equivalent to a million tons of dynamite.

Assume the average railroad car about 30 feet long, which can carry about 30 tons of load. A train carrying one million tons of dynamite would be about one million feet long.

$$1{,}000{,}000 \text{ ft.} \times \frac{1 \text{ mile}}{5{,}000 \text{ ft.}} = 200 \text{ miles.}$$

A train necessary to carry the chemical explosive equivalent to one bomb carried in the nuclear war head of a guided missile would be approximately 200 miles long (roughly the distance from New York to Baltimore). Where does this energy come from? Hydrogen is approximately the mass of a proton 1.0080 amu. Neutrons are about the same mass. Helium, which has a mass of 4.0026, contains two protons and two neutrons. However, $1.0080 \times 4 = 4.0320$ and $4.0320 - 4.0026 = .0294$ amu. The energy released in the hydrogen bomb must come from this missing mass.

$.0294/4.0320 \approx .007$. About .7 per cent of the mass disappears and becomes energy in a fusion reaction when hydrogen atoms are "squeezed" together to form helium.

The Twin Paradox

If a pair of twin brothers toss a coin to decide who is going to take a space voyage at close to the speed of light while the other one stays home, theory predicts that for the one who is traveling, time intervals would be dilated and he would age at a lower rate than his stay-at-home brother. They can argue, however, as follows. The brother who is at rest with respect to the earth "sees" the traveler moving away and coming back at near light speed and agrees that when the traveler returns he will be younger than he would have been if he stayed home. On the other hand, the traveler who is at rest with respect to his vehicle sees the earth moving away from him and returning at this high speed. He thus argues that it is the one who stayed on earth who will be younger when they again meet.

How then do they settle the argument? The traveler will first move out, decelerate, stop, and accelerate for the return trip. He will feel the forces required to decelerate and accelerate him. He knows he has been traveling. The twin who remains at home does not experience this acceleration and deceleration. He knows that he is the one who has remained at rest and will be the older when he rejoins his twin brother.

The advanced topics section of the PSSC physics has a chapter on relativity which has another interesting solution to the twin paradox. Suppose both agreed to broadcast their heartbeats as they moved away from each other. Since both are moving away from each other, both are affected by a doppler effect which makes the frequency smaller than it really is. When the traveler turns around, he is immediately heading back to the source of the heartbeat signal he is receiving and immediately knows he is going back or the source is approaching him because the frequency gets higher. Half the time of his trip he has experienced a low frequency and half the time he has experienced a high frequency. The twin who stays home, however, has a somewhat different experience.

He feels the low frequency as he and his brother separate, but when his traveling brother (the source of the frequency he is receiving) turns around he is not aware of it until the radio wave carrying the signal at the turn-around time reaches him. For him the symmetry of the trip is different. He has detected the lower doppler frequency longer than he has the higher frequency. When they compare notes on meeting again, they will know which was the traveler and which one remained at rest. There will be no surprise at finding which one appears older.

The Mossbauer Effect

A description of the Mossbauer Effect and its use in measuring in a laboratory, what could previously be seen only through cosmological observations, makes a very good vehicle for summarizing much of what had been discussed during the year in other contexts.

Consider Fe^{57}, an unstable isotope of iron which characteristically emits a gamma photon of a particular wave length as a nucleon in its nucleus falls to a lower energy level. Point out the correlation between energy level changes in the nucleus with emission of high energies and the energy changes of orbital electrons outside the nucleus with emission of relatively low energy photons. If a photon emitted from an excited Fe^{57} nucleus were to pass through other Fe^{57} atoms, it ought to be absorbed. Call attention to the Fraunhofer absorption and resulting dark line spectrum seen because photons from excited atoms on the surface of the sun must pass through the gases surrounding it. Remind your students of the discussions of resonance held earlier in the year in which you illustrated with tuned musical strings, tuning forks, and "tuned" pendulums that only those frequencies will be absorbed which would be the natural emitted frequency of the absorber if it were excited.

Here, though, you should point out that resonance cannot take place because of conservation of momentum and energy.

In order for a photon of wave length λ to be emitted, it must carry away a momentum of h/λ and the nucleus must recoil with a momentum equal to $-h/\lambda$. The emitted photon will, therefore, have to give up some of its energy to furnish the recoil energy of the nucleus. Its energy ($h\nu$) will be such that its frequency (ν) will be too small to be acceptable for resonance by an absorbing Fe_{57} atom. Even if the right frequency were emitted, it could not be absorbed since the absorbing atom would have to accept the momentum of the absorbed photon. In order to do that, the photon would need to carry a higher frequency than the natural radiation in order to supply both the excitation energy and the energy necessary for the nucleus to carry away its momentum.

It was Mossbauer's contribution to note that if the Fe^{57} was bound in a crystal lattice such that no single atom could move by itself, the mass of the crystal recoiling on emission or absorption of gamma photons would make it possible to conserve momentum with negligible effect on the energy of the photon passing between the emitter and absorber.

The truly remarkable property of this effect is its sensitivity. Suppose you had an absorbing crystal set up between an emitter and a geiger counter. You read the amount of radiation being received on the counter. If you then move the emitter toward or away from the receiver very slowly, the count goes up. (See Figure 21-6.)

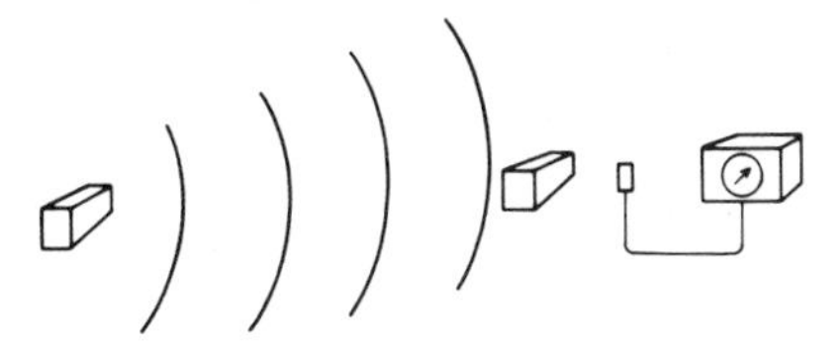

Figure 21-6

Moving the emitter causes a doppler shift in the frequency of the radiation arriving at the receiver which cannot resonate with the changed frequency and permits the radiation to pass through to be recorded by the counter.

This Mössbauer effect thus becomes an extremely sensitive tool. For instance, if $E = mc^2$, $m = E/c^2$ and since $F_g = GMm/r^2$ then $F_g = (GME/c^2)/r^2$, that is, if light photons carry mass, they should be affected by a strong gravitational field.

Caution: In teaching this section, it should be pointed out that the gravitational effect on light was predicted quantitatively by Einstein in his theory of general relativity which equates gravitational fields to acceleration. (If one were in a closed, windowless room and felt oneself being pulled toward the floor having weight, there would be no experiment one could perform which would differentiate between a large mass beneath you or the room and floor on which you stood being accelerated upwards.)

The relationship $E = mc^2$, which is derived from the theory of special relativity and which is based on the limiting speed of light, does (at least qualitatively) predict the influence of gravity on light.

Pictures taken of the sky at night and then of the same section of the sky during a solar eclipse show this gravitational influence. (See Figure 21-7.)

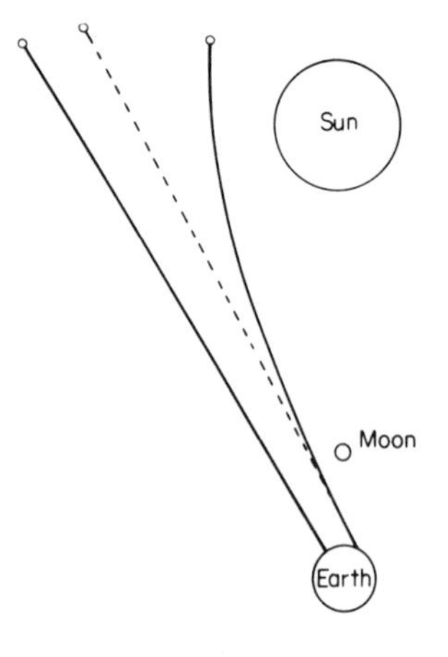

Figure 21-7

During a total solar eclipse, the light from one of the stars must pass closer to the massive sun in order to reach the camera on earth and is, therefore, deviated from a straight line path by the strong gravitational field of the sun.

It was further predicted that light leaving the surface of the sun must experience an increase in potential energy and, therefore, a loss in "Kinetic" energy. The only way light can show loss in energy is by increasing in wave length. This has become known as gravitational red shift.

What kind of experiment could be devised to check this out?

Pound and Rebke performed an experiment in 1960 in which the sensitive Mossbauer effect was used for detecting this gravitational effect on light photons.

Some of the old buildings at Harvard University have tall towers which were useful for the experiment I will now describe.

When an Fe^{57} crystal was attached to the roof of the tower, the gamma photons arriving at the floor of the tower should have lost potential energy and, therefore, experienced a "blue shift" (a shift toward a shorter wave length).

An absorbing crystal placed in front of a geiger counter can be used to detect this gravitational effect on the gamma photons emitted by the crystal on the top of the tower because the count recorded on the geiger counter went down when the absorber was moved slowly away from the emitting crystal. A doppler lengthening of the wave compensated for the gravitational shortening of the wave, and the absorbing crystal was able to absorb the gamma photons coming from above because the proper resonating frequency was now achieved.

It might be interesting to note that when the experiment was first tried it did not work, because of the difference in temperature at the top and bottom of the tower causing doppler shifts due to different average thermal motion of the atoms in the top and bottom crystals. It was only after the temperature of the emitting and receiving crystal was equalized that the experiment was successfully performed.

A useful discussion might be engendered by asking your students to design a temperature measuring instrument using the Mossbauer effect.

The relativistic time dilation, $\Delta t_r = \Delta t_0/\sqrt{1 - v^2/c^2}$, has also been demonstrated and measured in the laboratory with the Mossbauer effect.

Absorbing and emitting Mossbauer effect crystals are placed on a turn-table as shown in Figure 21-8. The reading on the geiger counter is read and then the turn-table is caused to revolve. The counter is then seen to receive a higher amount of radiation. Why?

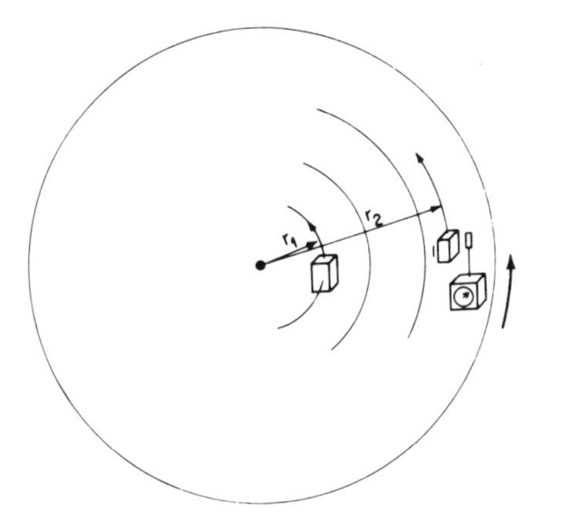

Figure 21-8

Certainly a motion transverse to the direction of propagation of waves can not lead to a change in wave length due to the classical doppler effect.

Note, however, that while both the emitter and receiver make a complete revolution in the same period (t), the emitter has only moved a distance

$2\pi r_1$ while the receiver has moved around the circumference of the larger circle $2\pi r_2$. The receiver is, therefore, moving faster than the emitter and for it, time is dilated. That is, the wave fronts from the emitter are arriving at a higher frequency than it can resonate with. The predicted change in frequency can be calculated by knowing the relativistic time dilation. (The interval of time, or the period, is the reciprocal of the frequency.)

The emitter is then moved toward the outside of the moving turn-table at the right speed to effect a doppler increase in frequency to compensate for this relativistic effect. The count is seen to go down, implying that resonance has once more been achieved.

CONCLUSION

This book has covered more material than most teachers will go through with their students in a year's work.

Yet, if at the end we seem not to have finished, perhaps that is how it should be. We have only begun the study of physics—the story is not complete and probably never will be.

While we have explored the universe from macrocosm to microcosm, we have shown that a few simple and fundamental basic concepts can be applied in various contexts.

If we have taught our students that man's truly awe-inspiring achievement is not his technology but his ability to understand, then we have reason to go on.

If we have awakened the interest of our students and fanned the spark of curiosity which will cause them to continue asking questions of nature, we can count ourselves successful.

Finally, if we interpret the meaning of a "liberal" education as making one more sensitive to the world in which we live, then we have made a contribution.

Index

L

M

S

T